Teaming with Microbes

Teaming
with
Microbes

A Gardener's Guide to the
Soil Food Web

Jeff Lowenfels & Wayne Lewis

Foreword by Elaine Ingham

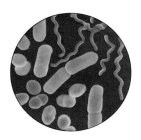

TIMBER PRESS

Disclaimer: The use of the scientific term "soil food web" is not intended to promote any commercial activity or to suggest any affiliation with a commercial entity.

Published in 2006 by
Timber Press, Inc.
The Haseltine Building
133 S.W. Second Avenue, Suite 450
Portland, Oregon 97204-3527, U.S.A.
www.timberpress.com
For contact information regarding editorial, marketing, sales, and distribution in the United Kingdom, see www.timberpress.co.uk.

Third printing in 2007

Printed in China

Library of Congress Cataloging-in-Publication Data

Lowenfels, Jeff.
    Teaming with microbes : a gardener's guide to the soil food web / by
Jeff Lowenfels and Wayne Lewis.
        p. cm.
    Includes bibliographical references (p.   ) and index.
    ISBN-13: 978-0-88192-777-1
    1. Soils. 2. Soil ecology. 3. Soil microbiology. 4. Food chains (Ecology)
I. Lewis, Wayne, 1942–   . II. Title.
S591.L59 2006
631.4—dc22

                                                            2005033502

A catalog record for this book is also available from the British Library.

*We dedicate this book to
our wives, Judith Hoersting and Carol Lewis,
who allowed us to team with microbes in the first place.
They married gardeners and ended up with
amateur microbiologists. They let compost tea brew
in the kitchen. They put up with fungi, bacteria,
nematodes, spiders, and worms. They remained silent
when we took the molasses from the cupboard.
They learned to turn compost piles and to
appreciate mushrooms in our lawns.*

Contents

Foreword

If you go down in the dirt today, you'd better not go alone!
For today's the day the nematodes have their picnic!

Sung to the tune of "The Teddy Bears' Picnic"

W HEN YOU ARE BORED looking at "soil" from urban lawns, making up words to popular songs is always good! Soil shouldn't be so boring, but urban landscapes mean dead dirt. It means being bent over a microscope for long hours looking at . . . nothing but inert particles. Boring. And so, we make up words to songs.

Real soil is active, alive, moving! Critters everywhere, doing interesting things! No need to invent new lyrics to old songs. No hours staring through a microscope looking at micrometer after micrometer of boring—nothing happening. Instead, after just a few seconds—movement, life, action!

Urban dwellers and other growers have been pouring toxic chemicals on their soils for years, without recognizing that those chemicals harm the very things that make soil healthy. Use of toxics to any extent creates a habitat for the "mafia" of the soil, an urban war zone, by killing off the normal flora and fauna that compete with the bad guys and keep them under control. Recent work strongly indicates that toxic chemicals destroy water quality, soil health, and the nutritional content of your food, because of the loss, eventually, of the beneficials in the soil. If toxic material was applied only once in your life, the bad situation we have today would not have developed, but typically with that first application, thousands of organisms that were beneficial to your plants were killed. A few bad guys were killed as well, but good guys are gone, and they don't come back as fast as the bad guys. Think about your neighborhood: who would come back faster if your neighborhood was turned into a chemical war zone? Opportunistic marauders and looters, that's who comes back in after disturbances. In the human world, we send in the National Guard, to hold the line against criminals. But in soil, the levels of inorganic fertilizer being used, or the constant applications of toxic pesticide sprayed, mean the National Guard of the soil has been killed, too. We have to purposefully restore the beneficial biology that has been lost.

Where will the new recruits come from? You have to add them—bacteria, fungi, protozoa, nematodes, earthworms, microarthropods—back to your soil. Roots of plants feed these beneficials, but to make sure that the beneficials get reestablished, care packages may need to be delivered. Soil Foodweb, Inc., helps people rapidly reestablish the biology that creates the foothold for health to come back into these systems; and this book describes these hardworking members of the front line of defense for your plants. Where do they live? Who are their families? How do you send in lunch packs, not toxics, to help the recruits along?

Win back your soil's health. Put nothing on your soil if you don't know what it will do to the life under your feet. If there is "no information" about how something impacts the life in your soil, or if the material has never been tested to determine what it does to the organisms in your soil, *don't use the material.* If you have already purchased the product, test it yourself.

Toxics are sometimes necessary to roust out a particularly bad infestation or disease, but toxics should be used as a last resort, not as your first response to a wilting plant. If you use toxics, then remember to replace the good guys, and send in some food, immediately.

Reestablishing the proper biology is critical. You may lose a few battles along the way. But persevere, and you can win. Think strategically: how can you help deliver troops, foods, medicines, and bandages to the front lines of the battle between beneficials and the diseases and pests in the most effective way? The directions, at least to the best of our knowledge, are in this book.

Most people have a great deal to learn when it comes to soil. You need the information that Jeff and Wayne have put together. They also make their "lessons" about soil health enjoyable! They present what could be deadly dull and boring in a way that is exciting and understandable. Instead of your having to work for years and years, staring through microscopes, as my colleagues and I have done in our efforts to understand soil biology, this book gives you an overview of what has been learned! The work of many scientists is brought together in this book, in a way that allows the complex story of life in the soil to be easily understood.

I hope you will join with us and help to learn how to return health to soil, and therefore, to the food you eat. The instructions are here.

Dr. Elaine Ingham, Ph.D.
President, Worldwide, Soil Foodweb, Inc.
www.soilfoodweb.com

Preface

WE WERE TYPICAL suburban gardeners. Each year, at the beginning of the growing season, we carpet bombed our lawns with a megadose of water-soluble, high-nitrogen fertilizer and watered like crazy; then we strafed their weeds with a popular broadleaf herbicide. Next, we attacked our vegetable gardens and flower beds with a bag or two of commercial fertilizer and leveled them with a rototiller until the soil, the color and texture of finely ground coffee, lay as smooth and level as the Bonneville Salt Flats. These things we did religiously, as did most of our neighbors. Once was never enough either. We continued to use chemical fertilizers throughout the season as if we were competing in the large-vegetable contest at the Alaska State Fair—and at the end of the season we rototilled again, for some inexplicable reason.

When necessary (and it often was), we would suit up into protective clothing—complete with rubber gloves and a face mask—and paint our birches to protect them from invading aphids by using some god-awful smelling stuff that listed ingredients no normal person could pronounce, assuming he or she took the time to read the incredibly small print on the chemical's label. Then we sprayed our spruce trees with something that smelled even worse—something so strong, one application lasted not one but two years. It was a good thing we did protect ourselves, as both spray products are now off the market, withdrawn as health hazards.

Don't misunderstand us. At the same time we were also practicing what we considered to be an "appropriate" measure of environmental responsibility and political correctness. We left the grass clippings on the lawn to decompose and tilled fallen leaves into the garden beds, and occasionally we let loose batches of lacewings, ladybird beetles, and praying mantids—our version of integrated pest management. We composted. We recycled our newspapers and aluminum cans. We fed the birds and allowed all manner of wildlife to wander in our yards. In our minds we were pretty organic and environmentally conscious (if not downright responsible). In short, we were like most home gardeners, maintaining just the right balance between better living with chemistry and at least *some* of Rachel Carson's teachings.

Besides, we were mostly using only water-soluble, high-nitrogen fertilizer. How bad could that be for the environment? It sure made the plants grow. And we really employed only one weed killer, albeit a nonselective, broadleaf one. Okay, we occasionally resorted to an insecticide too, but when we considered what was on the shelves of our favorite nurseries, these didn't amount to much in our minds. Surely we couldn't be causing harm when we were only trying to save a spruce, help a birch, or prevent noxious dandelions and chickweed from taking over the world?

Central to the way we cared for our gardens and yards was a notion shared by tens of millions of other gardeners and, until you finish this book, perhaps you as well: nitrogen from an organic source is the same as nitrogen from an inorganic one. Plants really didn't care if their nitrogen and other nutrients came from a blue powder you mixed with water or aged manure. It is all nitrogen to them.

Then one autumn, after the gardens were put to bed and we were settling in for the winter, looking for something to hold our horticultural interest for the cold months, a gardening friend e-mailed two stunning electron microscope pictures. The first showed in exquisite detail a nematode trapped by a single looped fungal strand, or hypha. Wow! This was quite a picture—a fungus taking out a nematode! We had never heard of, much less seen such a thing, and it started us wondering: how did the fungus kill its prey? What attracted the blind nematode to the rings of the fungus in the first place? How do the rings work?

The second image showed what appeared to be a similar nematode, only this one was unimpeded by fungal hyphae and had entered the tomato root. This photo raised its own questions. Why wasn't this nematode attacked, and where were the fungal hyphae that killed off the first nematode?

A foraging, root-eating nematode, trapped by a fungal hypha. Courtesy H. H. Triantaphyllou. Reprinted, with permission, from http://www.apsnet.org/, American Phytopathological Society, St. Paul, Minnesota.

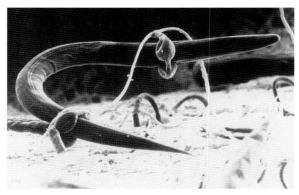

While researching the answers to these questions, we stumbled upon the work of Dr. Elaine Ingham, a soil microbiologist famous for her work with the life that resides in soil and, in particular, who eats whom in the soil world. Since some organisms eat from more than one food chain or are eaten by more than one type of predator, the chains are linked into webs—soil food webs. Ingham, an excellent teacher, became our guide to the whole world of complex communities in the soil. Through her we learned that the fungus in the first photograph was protecting the plant's roots; if that wasn't enough to make us stop and think, we learned the plant attracted the fungus to its roots in the first instance! And we also learned what killed the fungus that would have prevented the nematode from attacking the tomato root.

Naturally, we began to wonder what other heretofore-unseen things were going on down there in the soil. Might the world revealed to us by tools like the electron microscopes affect how we care for the plants in our gardens, yards, and lawns? We have all been dazzled by Hubble images of deep space, incomprehensibly far away, yet few of us have ever had the opportunity to marvel at the photographs produced by a scanning electron microscope (SEM), which provide a window to an equally unknown universe literally right under our feet.

We looked for answers, and soon realized that while we were out spreading fertilizer and rototilling our garden beds by rote, an ever-growing group of scientists around the world had been making discovery after discovery that put these practices into question. Many scientific disciplines—microbiology, bacteriology, mycology (the study of fungi), myrmecology (the study of ants), chemistry, agriculture—came together in recent decades to focus jointly on understanding the world of soil. Slowly, their findings about what goes on in the soil are being applied to commercial agriculture, silviculture, and viniculture. It is time we applied this science to things we grow in our home yards and gardens.

With no fungal hyphae barring the way, a nematode penetrates a tomato root to feed. Photograph by William Weryin and Richard Sayre, USDA-ARS.

Most gardeners are stuck in traditional horticultural land, a place where a blend of old wives' tales, anecdotal science, and slick commercial pitches designed to sell products dictates our seasonal activities. If there is any understanding of the underlying science of gardening, it is almost always limited to the soil's NPK chemistry and its physical structure. As you read these pages, you will learn how to use the biology in your soils—naturally or manipulated—to your and your plants' benefit. Since chemical fertilizers kill the soil microorganisms and chase away larger animals, the system we espouse is an organic one, free of chemicals. Chemicals, in fact, are what killed off the root-protecting fungal hyphae, giving our nematode friend access to the unprotected tomato root in the second photo.

By necessity, this book is divided into two sections. The first is an explanation of soil and the soil food web. There is no getting around it. You have to know the science before you can apply it. At least in this instance, the science is fascinating, even astonishing, and we try not to make a textbook out of it. The second section is the explanation of how to work the soil food web to your soils' advantage and to yours as a gardener.

What makes this book different from other texts on soil is our strong emphasis on the biology and microbiology of soils—relationships between soil and organisms in the soil and their impact on plants. We are not abandoning soil chemistry, pH, cation exchange, porosity, texture, and other ways to describe soil. Classic soil science is covered, but from the premise that it is the stage where the biology acts out its many dramas. After the players are introduced and their individual stories told, what evolves is a set of predictable outcomes from their interrelationships, or lack thereof. In the second half of the book, these outcomes are formed into a few simple rules, rules that we've applied in our yards and gardens, as have many of our neighbors in Alaska, where we initiated these new practices. So have others, throughout the Pacific Northwest in particular, but in other parts of the world as well. We think that learning about and then applying soil science (particularly the science of how various forms of life in the soil interrelate—the soil food web) has made us better gardeners. Once you are aware of and appreciate the beautiful synergisms between soil organisms, you will not only become a better gardener but a better steward of the earth. Home gardeners really have no business applying poisons, and yet apply them they do, to the food they grow and eat (and worse, feed to their families) and the lawns on which they play.

You might be tempted to skip right to the second part of this book, but we strongly discourage doing so. It is essential to know the science to really understand the rules. Sure, it requires a bit of effort (or the chapter on soil science does, anyway), but for too long, for too many gardeners, everything we needed to know came in a bottle or jar and all we had to do was mix with water and apply with a hose-end sprayer: instant cooking meets home gardening. Some hobby. Well, we want you to be thinking gardeners, not mindless consumers who react because a magazine or television ad says to do something. If you really want to be a good gardener, you need to understand what is going on in your soil.

So, here goes. We now know all nitrogen is not the same and that if you let the plants and the biology in the soil do their jobs, gardening becomes much easier and gardens much better. May your yard and your gardens grow to their natural glory. We know ours now do.

Part 1

The Basic Science

Electron microscope photograph of organic compost humus (brown), decaying plant material (green), and some mineral particles (purple and yellow), 25×. Image copyright Dennis Kunkel Microscopy, Inc.

Chapter 1

What Is the Soil Food Web and Why Should Gardeners Care?

IVEN ITS VITAL IMPORTANCE to our hobby, it is amazing that most of us don't venture beyond the understanding that good soil supports plant life, and poor soil doesn't. You've undoubtedly seen worms in good soil, and unless you habitually use pesticides, you should have come across other soil life: centipedes, springtails, ants, slugs, ladybird beetle larvae, and more. Most of this life is on the surface, in the first 4 inches (10 centimeters); some soil microbes have even been discovered living comfortably an incredible two miles beneath the surface. Good soil, however, is not just a few animals. Good soil is absolutely teeming with life, yet seldom does the realization that this is so engender a reaction of satisfaction.

In addition to all the living organisms you can see in garden soils (for example, there are up to 50 earthworms in a square foot [0.09 square meters] of good soil), there is a whole world of soil organisms that you cannot see unless you use sophisticated and expensive optics. Only then do the tiny, microscopic organisms—bacteria, fungi, protozoa, nematodes—appear, and in numbers that are nothing less than staggering. A mere teaspoon of good garden soil, as measured by microbial geneticists, contains a billion invisible bacteria, several yards of equally invisible fungal hyphae, several thousand protozoa, and a few dozen nematodes.

The common denominator of all soil life is that every organism needs energy to survive. While a few bacteria, known as chemosynthesizers, derive energy from sulfur, nitrogen, or even iron compounds, the rest have to eat something containing carbon in order to get the energy they need to sustain life. Carbon may come from organic material supplied by plants, waste products produced by other organisms, or the bodies of other organisms. The first order of business of all soil life is obtaining carbon to fuel metabolism—it is an eat-and-be-eaten world, in and on soil.

Do you remember the children's song about an old lady who accidentally swallowed a fly? She then swallows a spider ("that wriggled and jiggled and tickled inside her") to catch the fly, and then a bird to catch the spider, and so on, until she eats a horse and dies ("Of course!"). If you made a diagram of

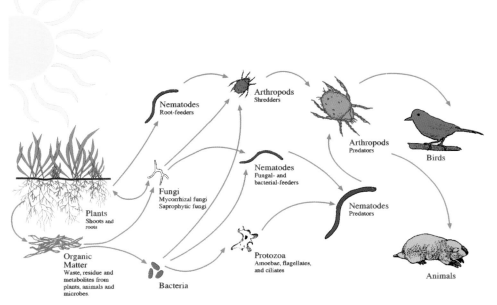

A soil food web, USDA-NRCS.

who was expected to eat whom, starting with the fly and ending with the improbable horse, you would have what is known as a food chain.

Most organisms eat more than one kind of prey, so if you make a diagram of who eats whom in and on the soil, the straight-line food chain instead becomes a series of food chains linked and cross-linked to each other, creating a web of food chains, or a soil food web. Each soil environment has a different set of organisms and thus a different soil food web.

This is the simple, graphical definition of a soil food web, though as you can imagine, this and other diagrams represent complex and highly organized sets of interactions, relationships, and chemical and physical processes. The story each tells, however, is a simple one and always starts with the plant.

Plants are in control

Most gardeners think of plants as only taking up nutrients through root systems and feeding the leaves. Few realize that a great deal of the energy that results from photosynthesis in the leaves is actually used by plants to produce chemicals they secrete through their roots. These secretions are known as exudates. A good analogy is perspiration, a human's exudate.

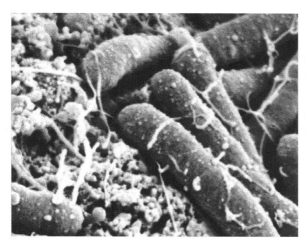

The rhizosphere is an area of interaction between the surface of a plant root and the area surrounding it. Bacteria and other micro-organisms as well as soil debris fill the area. 10,000×. Photograph by Sandra Silvers, USDA-ARS.

Root exudates are in the form of carbohydrates (including sugars) and proteins. Amazingly, their presence wakes up, attracts, and grows specific beneficial bacteria and fungi living in the soil that subsist on these exudates and the cellular material sloughed off as the plant's root tips grow. All this secretion of exudates and sloughing-off of cells takes place in the rhizosphere, a zone immediately around the roots, extending out about a tenth of an inch, or a couple of millimeters (1 millimeter = $\frac{1}{25}$ inch). The rhizosphere, which can look like a jelly or jam under the electron microscope, contains a constantly changing mix of soil organisms, including bacteria, fungi, nematodes, protozoa, and even larger organisms. All this "life" competes for the exudates in the rhizosphere, or its water or mineral content.

At the bottom of the soil food web are bacteria and fungi, which are attracted to and consume plant root exudates. In turn, they attract and are eaten by bigger microbes, specifically nematodes and protozoa (remember the amoebae, paramecia, flagellates, and ciliates you should have studied in biology?), who eat bacteria and fungi (primarily for carbon) to fuel their metabolic functions. Anything they don't need is excreted as wastes, which plant roots are readily able to absorb as nutrients. How convenient that this production of plant nutrients takes place right in the rhizosphere, the site of root-nutrient absorption.

At the center of any viable soil food web are plants. Plants control the food web for their own benefit, an amazing fact that is too little understood and surely not appreciated by gardeners who are constantly interfering with Nature's system. Studies indicate that individual plants can control the numbers and the different kinds of fungi and bacteria attracted to the rhizosphere by the exudates they produce. During different times of the growing season, populations of rhizosphere bacteria and fungi wax and wane, depending on the nutrient needs of the plant and the exudates it produces.

Soil bacteria and fungi are like small bags of fertilizer, retaining in their bodies nitrogen and other nutrients they gain from root exudates and other organic matter (such as those sloughed-off root-tip cells). Carrying on the analogy, soil protozoa and nematodes act as "fertilizer spreaders" by releasing the nutrients locked up in the bacteria and fungi "fertilizer bags." The nematodes and protozoa in the soil come along and eat the bacteria and fungi in the rhizosphere. They digest what they need to survive and excrete excess carbon and other nutrients as waste.

Left to their own devices, then, plants produce exudates that attract fungi and bacteria (and, ultimately, nematodes and protozoa); their survival depends on the interplay between these microbes. It is a completely natural system, the very same one that has fueled plants since they evolved. Soil life provides the nutrients needed for plant life, and plants initiate and fuel the cycle by producing exudates.

Soil life creates soil structure

The protozoa and nematodes that feasted on the fungi and bacteria attracted by plant exudates are in turn eaten by arthropods (animals with segmented bodies, jointed appendages, and a hard outer covering called an exoskeleton).

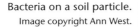
Bacteria on a soil particle.
Image copyright Ann West.

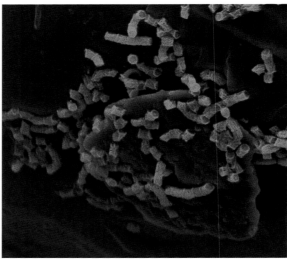

Insects, spiders, even shrimp and lobsters are arthropods. Soil arthropods eat each other and themselves are the food of snakes, birds, moles, and other animals. Simply put, the soil is one big fast-food restaurant. In the course of all this eating, members of a soil food web move about in search of prey or protection, and while they do, they have an impact on the soil.

Bacteria are so small they need to stick to things, or they will wash away; to attach themselves, they produce a slime, the secondary result of which is that individual soil particles are bound together (if the concept is hard to grasp, think of the plaque produced overnight in your mouth, which enables mouth bacteria to stick to your teeth). Fungal hyphae, too, travel through soil particles, sticking to them and binding them together, thread-like, into aggregates.

Worms, together with insect larvae and moles and other burrowing animals, move through the soil in search of food and protection, creating pathways that allow air and water to enter and leave the soil. Even microscopic fungi can help in this regard (see chapter 4). The soil food web, then, in addition to providing nutrients to roots in the rhizosphere, also helps create soil structure: the activities of its members bind soil particles together even as they provide for the passage of air and water through the soil.

Soil life produces soil nutrients

When any member of a soil food web dies, it becomes fodder for other members of the community. The nutrients in these bodies are passed on to other members of the community. A larger predator may eat them alive, or they may be decayed after they die. One way or the other, fungi and bacteria get involved, be it decaying the organism directly or working on the dung of the successful eater. It makes no difference. Nutrients are preserved and eventually are retained in the bodies of even the smallest fungi and bacteria. When these are in the rhizosphere, they release nutrients in plant-available form when they, in turn, are consumed or die.

Without this system, most important nutrients would drain from soil. Instead, they are retained in the bodies of soil life. Here is the gardener's truth: when you apply a chemical fertilizer, a tiny bit hits the rhizosphere, where it is absorbed, but most of it continues to drain through soil until it hits the water table. Not so with the nutrients locked up inside soil organisms, a state known as immobilization; these nutrients are eventually released as wastes, or mineralized. And when the plants themselves die and are allowed to decay, the nutrients they retained are again immobilized in the fungi and bacteria that consume them.

The nutrient supply in the soil is influenced by soil life in other ways. For example, worms pull organic matter into the soil, where it is shredded by beetles and the larvae of other insects, opening it up for fungal and bacterial decay. This worm activity provides yet more nutrients for the soil community.

Healthy soil food webs control disease

A healthy food web is one that is not being destroyed by pathogenic and disease-causing organisms. Not all soil organisms are beneficial, after all. As gardeners you know that pathogenic soil bacteria and fungi cause many plant diseases. Healthy soil food webs not only have tremendous numbers of individual organisms but a great diversity of organisms. Remember that teaspoon of good garden soil? Perhaps 20,000 to 30,000 different species make up its billion bacteria—a healthy population in numbers *and* diversity.

A large and diverse community controls troublemakers. A good analogy is a thief in a crowded market: if there are enough people around, they will catch or even stop the thief (and it is in their self-interest to do so). If the market is deserted, however, the thief will be successful, just as he will be if he is stronger, faster, or in some other way better adapted than those that would be in pursuit.

In the soil food web world, the good guys don't usually catch thieves (though it happens: witness the hapless nematode that started this all for us); rather, they compete with them for exudates and other nutrients, air, water, and even space. If the soil food web is a healthy one, this competition keeps the pathogens in check; they may even be outcompeted to their death.

Just as important, every member of the soil food web has its place in the soil community. Each, be it on the surface or subsurface, plays a specific role. Elimination of even just one group can drastically alter a soil community. Birds participate by spreading protozoa carried on their feet or dropping a worm taken from one area into another. Too many cats, and things will change. Dung from mammals provides nutrients for beetles in the soil. Kill the mammals, or eliminate their habitat or food source (which amounts to the same thing), and you won't have as many beetles. It works in the reverse as well. A healthy soil food web won't allow one set of members to get so strong as to destroy the web. If there are too many nematodes and protozoa, the bacteria and fungi on which they prey are in trouble and, ultimately, so are the plants in the area.

And there are other benefits. The nets or webs fungi form around roots act as physical barriers to invasion and protect plants from pathogenic fungi and bacteria. Bacteria coat surfaces so thoroughly, there is no room for others to attach themselves. If something impacts these fungi or bacteria and their numbers drop or they disappear, the plant can easily be attacked.

Special soil fungi, called mycorrhizal fungi, establish themselves in a symbiotic relationship with roots, providing them not only with physical protection but with nutrient delivery as well. In return for exudates, these fungi provide water, phosphorus, and other necessary plant nutrients. Soil food web populations must be in balance, or these fungi are eaten and the plant suffers.

Bacteria produce exudates of their own, and the slime they use to attach to surfaces traps pathogens. Sometimes, bacteria work in conjunction with fungi to form protective layers, not only around roots in the rhizosphere but on an equivalent area around leaf surfaces, the phyllosphere. Leaves produce exudates that attract microorganisms in exactly the same way roots do; these act as a barrier to invasion, preventing disease-causing organisms from entering the plant's system.

Some fungi and bacteria produce inhibitory compounds, things like vitamins and antibiotics, which help maintain or improve plant health; penicillin and streptomycin, for example, are produced by a soil-borne fungus and a soil-borne bacterium, respectively.

All nitrogen is not the same

Ultimately, from the plant's perspective anyhow, the role of the soil food web is to cycle down nutrients until they become temporarily immobilized in the bodies of bacteria and fungi and then mineralized. The most important of these nutrients is nitrogen—the basic building block of amino acids and, therefore, life. The biomass of fungi and bacteria (that is, the total amount of each in the soil) determines, for the most part, the amount of nitrogen that is readily available for plant use.

It wasn't until the 1980s that soil scientists could accurately measure the amount of bacteria and fungi in soils. Dr. Elaine Ingham at Oregon State University along with others started publishing research that showed the ratio of these two organisms in various types of soil. In general, the least disturbed soils (those that supported old growth timber) had far more fungi than bacteria, while disturbed soils (rototilled soil, for example) had far more bacteria than fungi. These and later studies show that agricultural soils have a fungal to bacterial biomass (F:B ratio) of 1:1 or less, while forest soils have ten times or more fungi than bacteria.

Ingham and some of her graduate students at OSU also noticed a correlation between plants and their preference for soils that were fungally dominated versus those that were bacterially dominated or neutral. Since the path from bacterial to fungal domination in soils follows the general course of plant succession, it became easy to predict what type of soil particular plants preferred by noting where they came from. In general, perennials, trees, and shrubs prefer fungally dominated soils, while annuals, grasses, and vegetables prefer soils dominated by bacteria.

One implication of these findings, for the gardener, has to do with the nitrogen in bacteria and fungi. Remember, this is what the soil food web means to a plant: when these organisms are eaten, some of the nitrogen is retained by the eater, but much of it is released as waste in the form of plant-available ammonium (NH_4). Depending on the soil environment, this can either remain as ammonium or be converted into nitrate (NO_3) by special bacteria. When does this conversion occur? When ammonium is released in soils that are dominated by bacteria. This is because such soils generally have an alkaline pH (thanks to bacterial bioslime), which encourages the nitrogen-fixing bacteria to thrive. The acids produced by fungi, as they begin to dominate, lower the pH and greatly reduce the amount of these bacteria. In fungally dominated soils, much of the nitrogen remains in ammonium form.

Ah, here is the rub: chemical fertilizers provide plants with nitrogen, but most do so in the form of nitrates (NO_3). An understanding of the soil food web makes it clear, however, that plants that prefer fungally dominated soils ultimately won't flourish on a diet of nitrates. Knowing this can make a great deal of difference in the way you manage your gardens and yard. If you can cause either fungi or bacteria to dominate, or provide an equal mix (and you can—just how is explained in Part 2) , then plants can get the kind of nitrogen they prefer, without chemicals, and thrive.

Negative impacts on the soil food web

Chemical fertilizers negatively impact the soil food web by killing off entire portions of it. What gardener hasn't seen what table salt does to a slug? Fertilizers are salts; they suck the water out of the bacteria, fungi, protozoa, and nematodes in the soil. Since these microbes are at the very foundation of the soil food web nutrient system, you have to keep adding fertilizer once you start using it regularly. The microbiology is missing and not there to do its job, feeding the plants.

It makes sense that once the bacteria, fungi, nematodes, and protozoa are gone, other members of the food web disappear as well. Earthworms, for example, lacking food and irritated by the synthetic nitrates in soluble nitrogen fertilizers, move out. Since they are major shredders of organic material, their absence is a great loss. Without the activity and diversity of a healthy food web, you not only impact the nutrient system but all the other things a healthy soil food web brings. Soil structure deteriorates, watering can become problematic, pathogens and pests establish themselves and, worst of all, gardening becomes a lot more work than it needs to be.

If the salt-based chemical fertilizers don't kill portions of the soil food web, rototilling will. This gardening rite of spring breaks up fungal hyphae, decimates worms, and rips and crushes arthropods. It destroys soil structure and eventually saps soil of necessary air. Again, this means more work for you in the end. Air pollution, pesticides, fungicides, and herbicides, too, kill off important members of the food web community or "chase" them away. Any chain is only as strong as its weakest link: if there is a gap in the soil food web, the system will break down and stop functioning properly.

Healthy soil food webs benefit you and your plants

Why should a gardener be knowledgeable about how soils and soil food webs work? Because then you can manage them so they work for you and your plants. By using techniques that employ soil food web science as you garden, you can at least reduce and at best eliminate the need for fertilizers, herbicides, fungicides, and pesticides (and a lot of accompanying work). You can improve degraded soils and return them to usefulness. Soils will retain nutrients in the bodies of soil food web organisms instead of letting them leach out to God knows where. Your plants will be getting nutrients in the form each particular plant wants and needs so they will be less stressed. You will have natural disease prevention, protection, and suppression. Your soils will hold more water.

The organisms in the soil food web will do most of the work of maintaining plant health. Billions of living organisms will be continuously at work throughout the year, doing the heavy chores, providing nutrients to plants, building defense systems against pests and diseases, loosening soil and increasing drainage, providing necessary pathways for oxygen and carbon dioxide. You won't have to do these things yourself.

Gardening with the soil food web is easy, but you must get the life back in your soils. First, however, you have to know something about the soil in which the soil food web operates; second, you need to know what each of the key members of the food web community does. Both these concerns are taken up in the rest of Part 1.

Chapter 2
Classic Soil Science

T HIS WOULD BE A GOOD TIME to go outside and get a few handfuls of soil from different places in your yard. Take a good, close look at the soil. Smell it. Grind some between your fingers. Compare the samples for differences and similarities. When you repeat these observations after you read this chapter, you will have a different perspective of what is in your hands.

The typical gardener knows very little about soil and why it matters. To us, however, soil is the house in which all the organisms of the soil food web live. It is the stage for the actors that interest us. You simply have to know something about the physical nature of soil if you are to understand the biology that inhabits it and how to use this biology to become a better gardener. After all, an acre of good garden soil teems with life, containing several pounds (about 1 kilogram) of small mammals; 133 pounds of protozoa; 900 pounds each of earthworms, arthropods, and algae; 2000 pounds of bacteria; and 2400 pounds of fungi.

Most of us, if we want things to grow better, simply replace soil that is poor in quality with good soil. Experienced gardeners know good soil when they see it: coffee-colored, rich in organic matter, able to hold water yet still drain when there is too much around. And it smells good. Poor soil is pale, compacted, drains either too well and won't retain any water or holds too much water, sometimes even becoming anaerobic. It can smell bad. If you are going to use the soil food web, however, you really need to know more. Where does soil come from? What are its components? How can we agree to describe it, and how can we measure its characteristics? This knowledge will help you adjust your soils, for what determines really good soil, in the end, is what you wish to grow in it: good soil must be able to maintain a soil food web compatible with the plants it supports. Trust us—in the end, you will be glad you know a little something more about soil, something beyond its color and smell.

What is soil, really?

Technically, soil is all the loose, unconsolidated, mineral and organic matter in the upper layer of the earth's crust. The standard comparison uses an apple to represent the earth. Carve off approximately 75% of the skin, which represents all the water, and another 15%, which represents deserts or mountains—land too hot, too cold, too wet, or too steep to be usable for growing plants. The 10% that remains represents all the earth's soil—soils with the necessary physical, chemical, and biological properties to support plant life. When we take into account the footprints of cities, roads, and other man-made infrastructure (these, incidentally, usually are sited on some of the very best soils), the surface area of usable soil is further reduced.

For the moment, the thing that concerns us is the tiny strip of apple skin that represents the soil in our gardens and yards. How did it get there? What is it? Why does it support plant growth?

Weathering

Your yard's soil is in large part a product of weathering. Weathering is the sum impact of all the natural forces that decay rocks. These forces can be physical, chemical, or biological.

To begin, the mere action of wind, rain, snow, sun, and cold (along with glacial grinding, bumps along river beds, scrapes against other rocks, and rolls in ocean waves and stream currents) physically breaks rocks down into tiny mineral particles and starts the process of soil formation. Water freezes in rock cracks and crevices and expands, increasing its volume by 9% (and exerting a force of about 2000 pounds per square inch) as it turns to ice. Hot weather causes the surfaces of rock to expand, while the inner rock, just a millimeter away, remains cool and stable. As the outer layer pulls away, cracks form, and the surface peels off into smaller particles.

Chemical weathering dissolves rock by breaking the molecular bonds that hold it together through exposure to water, oxygen, and carbon dioxide. Some materials in rock go into solution, causing the rock to lose structural stability and making it more susceptible to physical weathering (think of a sugar cube dropped into a cup of tea and then stirred). Fungi and bacteria also contribute to chemical weathering by producing chemicals as they decay their food (fungi produce acids, and bacteria alkaline substances); besides carbon dioxide, microbes produce ammonia and nitric acids, which act as solvents. Rock material is broken down into simpler elements. Although there are almost 90 different chemical elements in soil, only eight constitute the majority: oxygen, silicon, aluminum, iron, magnesium, calcium, sodium, and potassium. All have an electric charge on a molecular level, and in different combinations these form electrically charged molecules that combine to form different minerals.

The acids produced by the yellow lichen on this rock are slowly contributing to its conversion into soil. Photograph by Dave Powell, USDA Forest Service, www.forestryimages.org.

Biological activity, too, causes weathering. Mosses and lichens (or, more precisely, the fungi in them) attach themselves to rocks and produce acids and chelating agents that dissolve little bits of rock to use as nutrients, resulting in small fissures that fill up with water. Freeze and thaw cycles further break apart the parent material, and the roots of larger plants penetrate crevices and widen them, forcing rocks apart.

Organic matter

Weathering breaks rock down into mineral components of one sort or another. Soil, however, needs to be able to support plant life—and that requires more than just minerals. On average, good garden soil is 45% mineral in nature and 5% organic matter, built up as organisms above and in it go about their daily business. As plants and animals on the surface die and are decayed by bacteria and fungi, they are ultimately converted into humus, a carbon-rich, coffee-colored, organic material. Think of the end product of composting. This valuable material is humus.

Humus consists of very long, hard-to-break chains of carbon molecules with a large surface area; these surfaces carry electrical charges, which attract and hold mineral particles. What's more, the molecular structure of the long chains resembles a sponge—lots of nooks and crannies that serve as veritable condominiums for soil microbes. Once you've added humus and other organic matter, such as dead plant matter and insect bodies, to weathered minerals, you have a soil almost capable of supporting trees, shrubs, lawns, and gardens—but not quite.

Air and water

Minerals and humus make up the solid phase of soil, but plants require oxygen and water—the gaseous and liquid phases—as well. The voids between individual mineral and organic particles are filled by air or water (and sometimes both).

Humus has a rich, coffee color and is full of organic material. This handful is about 55% organic matter. Courtesy Alaska Humus Company, www.alaskahumus.com.

Water moves between soil pore spaces in one of two ways: by the pull of gravity or by the pull of individual water molecules on each other, or capillary action. Gravitational water moves freely through soils. Picture water being poured into a jar of gravel: gravity pulls the water to the bottom as the jar fills up. Large pores promote the flow of gravitational water. As the water fills the pores, it displaces and pushes out the air in front of it. When the water flows through, it allows a new supply of air to move in. When gravitational water hits roots, which act like sponges, it is absorbed.

Smaller soil pore spaces contain a film of capillary water that is not influenced by gravity and is actually left behind after gravitational water passes through. The liquid is bonded together by the attraction of its molecules for each other (a force known as cohesion, but let's not complicate things) and to surrounding soil surfaces (a force known as adhesion). This creates a surface tension, causing the water to form a thick film on the particle surfaces. Capillary water can "flow" uphill. It is available to plant roots after gravitational water has passed by and as such is a major source of water for plants.

Hydroscopic water is a thinner film of water, only a few molecules thick, which, like capillary water, is attached to extremely small soil particles by virtue of electrical properties. This film is so thin that the bonds between water molecules and soil particles are concentrated and extremely hard to break. Roots cannot absorb it, therefore, but this film of water is critical to the ability of many microbes to live and travel. Even when conditions are dry, the soil particulate surface holds some hydroscopic water; it is impossible to remove it from soil without applying lots of heat and actually boiling it off.

Just about half the pore spaces in good soil are filled with water. The other half are filled with air. Water movement pushes stale air out and sucks in air from the surface, so adding water means an exchange of air occurs, which is important. If a healthy soil food web is in place, the metabolic activity of soil organisms uses the oxygen and creates carbon dioxide. The presence of carbon dioxide is a good sign that the soil contains life; however, the carbon dioxide must be exchanged with fresh air to keep life going.

In some soils, the pore spaces are cut off in lots of places, and air is not exchanged when water flows. In fact, water may not flow at all. These soils have very poor porosity—that is, they lack adequate space between the soil particles. All the oxygen in the soil can be used up by aerobic metabolic activities, resulting in oxygen-less, anaerobic conditions. Organisms that can live in such conditions often produce alcohols and other substances that kill plant root cells.

Soil profiles and horizons

Soils are exposed nonstop to the forces of weathering. Rain, for example, will cause some soil minerals and organic matter to leach out as the water moves down through the soil. This material may hit an impervious barrier and become concentrated in a certain zone or layer. The size of particles may cause a particular material to be concentrated or be filtered. Eventually, over time, distinct layers and zones of different material are formed. These can be seen, like the rock strata in the Grand Canyon's walls, as you dig down through the soil. A soil profile is a map of these layers, or horizons.

Soil scientists have attached a letter or combination of letters (and even numbers) to each horizon that appears in any typical soil profile. For the gardener (thankfully), the top horizons, the O and A, are really the only ones that count. The Oi horizon contains organic material that can still be specifically identified (with a bit of training that's beyond the scope of this book); this is fibric soil. The Oe horizon has experienced more decay, and while the materials are identifiable as plant matter, you cannot tell which specific plants are involved, even with training; this is humic soil. Finally, the Oa horizon is where the organic material has decomposed so much that you cannot tell its origin. It could be from plants or from animal matter. This is sapric soil. All this is somewhat useful information if you want to know if your soils will create more decay by-products (like nitrogen) because the process that converts the soil to humus isn't complete; or if your soil has been decayed to the point where it basically just houses microbes that cause decay.

The A horizon lies under the O horizon. Here humus particles accumulate as water runs through the O horizon above and pulls organic particles downward. Water flowing through this horizon carries lots of dissolved and suspended materials. This A horizon has the highest organic matter content of any of the soil horizons and the highest biological activity. This is where the roots grow.

Several other soil horizons follow and, eventually, bedrock. You would need a backhoe to trench through all the horizons under your yard, something that is clearly not worth the effort. Often one or more horizons are missing, worn or transported away by weathering forces, and just as often it is too hard to see any distinction between layers.

The important thing is to make sure your gardens and yard have good soil—the proper mixture of minerals, organic matter, air, and water—in the top layers, in the area where plants grow. If not, you will have to add to what you do have or replace it entirely.

Soil color

Color can be an easy indicator of what is in your soil, as soil color is sometimes dependent on the soil's specific mineral and organic components. Weathering, oxidation, reduction actions of iron and manganese minerals, and the biochemistry of the decomposition of organic matter are the primary factors influencing soil color.

Organic components in soil are very strong coloring agents and produce dark soils; these can accumulate or can dissolve and coat other particles of soil with black color. When iron is a component of soil, it rusts, and soil particles are coated with red and yellowish tints. When manganese oxide is a major component of soil, its particles take on a purple-black hue. The presence of these colors usually indicates good drainage and aeration.

Gray soils can indicate a lack of organic material. They also often indicate anaerobic conditions because the microbes that survive in such conditions often use the iron in the soil, rendering it colorless in the process. Similarly, magnesium is reduced to colorless compounds by other types of anaerobic soil microbes.

Soil scientists use color charts to identify, compare, and describe soil conditions. For the gardener, however, color plays less of a role. For us, good soil is the color of dark coffee—again, mostly because of its organic components.

Soil texture

Soil scientists describe the size of soil particles in terms of texture. There are three categories of soil texture: sand, silt, and clay. All soil has a specific texture that enables one to judge its propensity to support a healthy soil food web and thus healthy plants.

Soil texture has nothing to do with composition. If you think the term "sand" applies only to quartz particles, for example, you would be wrong. True, most sand particles are mineral quartz, but all sorts of rock can be weathered into sand: silicates, feldspars (potassium-aluminum silicate, sodium-aluminum silicate, and calcium-aluminum silicate), iron, and gypsum (calcium sulfate). If the sand comes from ground-up coral reefs, it is limestone. Most silt particles, too, are mineral quartz (only they are much smaller in size than those found in sandy soils), and silts can have the same non-quartz constituents as sand. Clays, on the other hand, are made up of an entirely different group of minerals, hydrous aluminum silicates, with other elements, such as magnesium or iron, occasionally substituting for some of the aluminum.

So, the key point for the gardener is that texture has to do with size of particles only, not the composition of these particles. What size particles, then, constitute sand, slit, and clay?

Start with sand. You've undoubtedly been to a beach and know that sand particles can be seen with the naked eye. They range in size from 0.0625 to 2 millimeters in diameter. Anything much bigger has far too much space between individual particles to be of any use to gardeners except as gravel for a path. Sand particles are just small enough to hold some water when aggregated, but most of it is gravitational water and readily drains out, leaving lots of air and only a little capillary water. Moreover, the particles of sand are big enough to be influenced by gravity, and they quickly settle to the bottom when mixed in water. As to texture, soils with large proportions of sand in them are gritty when ground between the fingers.

Next in texture size is silt. Sand particles can be seen with the naked eye, but you will need a microscope to see individual silt particles. Like sand, these consist of weathered rock, only much, much smaller in size—between 0.004 and 0.0625 millimeter in diameter. The pore spaces between silt particles are much smaller and hold a lot more capillary water than sand does. Like sand, particles of silt are also influenced by gravity and will settle out when put in water. The texture of silt when rubbed between fingers is that of flour.

Clays are formed during intense hydrothermal activity or by chemical action, that of carbonic acid weathering silicate-bearing rocks. Clay particles are readily distinguished from silt, but this time an electron microscope is

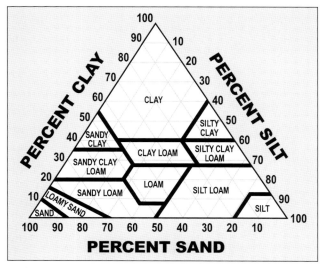

A diagram of soil textures. Courtesy Tom Hoffman Graphic Design.

needed—these particles are that small, the smallest that make up soil, 0.004 millimeter across or less. Clay particles are "plastic" and somewhat slippery when rubbed between fingers. This is because clay particles absorb and hold lots of water, which is why they are known as hydrous silicate compounds. Besides silicon, they contain water and often aluminum, magnesium, and iron as well.

For comparison, let's put things into a more familiar perspective. If a clay particle were the size of a marigold seed, a silt particle would be a large radish, and a sand grain would be a large garden wheelbarrow. Another way to look at soil texture is to visualize a gram (about a teaspoon's worth) of sand, spread out one-particle deep; this would cover an area about the size of a silver dollar. If you were to spread an equal amount of clay one-particle thick, you would need a basketball court—and some of the stands surrounding it, at that.

What difference does texture make? The size of the particles has everything to do with their surface area and the surface area of the pore spaces between individual particles. Clay has tremendous surface area compared to sand. Silt is in between. Clay has smaller pore spaces between particles, but many more pore spaces in total, so the surface area of the pore spaces in clay is greater than silt, which is greater than sand. Incidentally, organic matter, usually in the form of humus, is comprised of very minute particles that, like clay, have lots of surface area to which plant nutrients attach, thus preventing them from leaching out. Humus also holds capillary water.

All soils have different textures, but any can be put into a specific category, depending on how much sand-, silt-, and clay-sized grains they contain. The ideal garden soil is loam, a mixture with relatively equal parts of sand, silt, and clay. Loam has the surface area of silt and clay, to hold nutrients and water, and the pore space of sand, to aid drainage and help pull in air.

Sample your soils

Good garden soil contains 30 to 50% sand, 30 to 50% silt, and 20 to 30% clay, with 5 to 10% organic matter. You can find out how close your soils come to this ideal, loam. All it takes is a quart jar, two cups of water, and a tablespoon of a water softener, such as Calgon liquid. You will also need soil from the top 12 inches (30 centimeters) of the areas you want tested, be it your vegetable garden, flower bed, or lawn.

Mix each soil sample with two cups of water and a tablespoon of water softener. Put it in the jar, close the jar, and shake it vigorously, so that all the particles become suspended in the water. Then put the jar down and let things settle. After a couple of minutes, any sand particles in your soil will have settled out. It takes a few hours for the smallest silt particles to settle on top of this sand. Much of the smallest clay-sized particles will actually stay in suspension for up to a day. Organics in the soil will float to the top and remain there for an even longer period.

Wait 24 hours and then measure the thickness of each of the layers with a ruler. To determine the percentages of each, divide the depth or thickness of each layer by the total depth of all three layers and then multiply the answer by 100. Once you know what percentages of each material are in your soil, you can begin to physically change it if need be. How to do this is discussed in the second half of the book.

Soil structure

Individual particulate size, or texture, is obviously an important characteristic of soils, but so is the actual shape these particles take when grouped together. This shape, or soil structure, depends on both the soil's physical and chemical properties. Factors that influence soil structure are particle orientation, amount of clay and humus, shrinking and swelling due to weather (wetting and drying as well as freezing and thawing), root forces, biological influences (worms and small animals), and human activity. Soil structure types, or peds, fall into several distinct categories.

When you look at your garden soils, you don't see individual particles but rather aggregates of these particles. The biology in the soil produces the glues that bind individual soil particles into aggregates. As they go about their day-to-day business, bacteria, fungi, and worms produce polysaccharides, sticky carbohydrates that act like glues, binding individual mineral and humic particles together into aggregates.

Let's start with bacteria. The slime they produce allows them to stick to particles as well as to each other. Colonies are formed, and these too stick together, as do the particles to which the bacteria are attached. Fungi also help create soil aggregates. A group of common soil fungi, in the order Glomales, produces a sticky protein called glomalin. As the fungal strands, or hyphae, grow through soil pores, glomalin coats soil particles like super glue, sticking these particles together into aggregates or clumps. These aggregates change the soil pore space, making it easier for the soil to hold capillary water and soluble nutrients and recycle them slowly to plants.

Worms process soil particles in search of food. Individual particles of minerals and organics are ingested and ultimately excreted as aggregates; these are so large, they are readily identified as worm castings. Consider, too, the impact

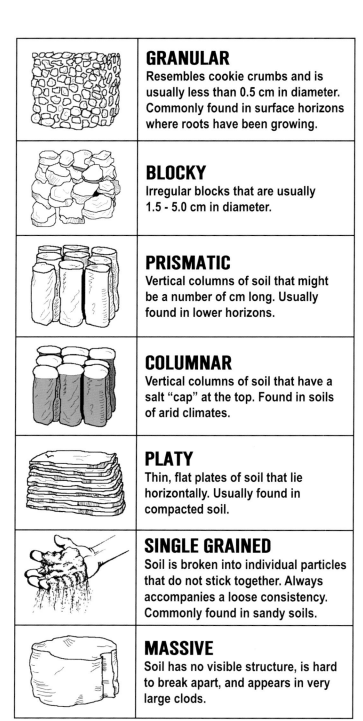

GRANULAR
Resembles cookie crumbs and is usually less than 0.5 cm in diameter. Commonly found in surface horizons where roots have been growing.

BLOCKY
Irregular blocks that are usually 1.5 - 5.0 cm in diameter.

PRISMATIC
Vertical columns of soil that might be a number of cm long. Usually found in lower horizons.

COLUMNAR
Vertical columns of soil that have a salt "cap" at the top. Found in soils of arid climates.

PLATY
Thin, flat plates of soil that lie horizontally. Usually found in compacted soil.

SINGLE GRAINED
Soil is broken into individual particles that do not stick together. Always accompanies a loose consistency. Commonly found in sandy soils.

MASSIVE
Soil has no visible structure, is hard to break apart, and appears in very large clods.

Soil structure peds. Courtesy Tom Hoffman Graphic Design.

of soil organisms as they make their way through the soil. Each group of animals has various body widths. As they move, they create spaces in and between soil particles and aggregates. By way of comparison, imagine that a bacterium 1 micrometer in diameter (1 micrometer = $^1/_{25,000}$ inch) is the width of a piece of spaghetti. Fungal bodies are generally wider, 3 to 5 micrometers. Nematodes (5 to 100 micrometers on average) would be the size of a pencil, perhaps even one of those thick ones; and protozoa (10 to 100 micrometers) would be the diameter of an American-style hot dog. Continuing to use our scale, soil mites and springtails, at 100 micrometers to 5 millimeters, would have the diameter of a good-sized tree. Beetles, earthworms, and spiders (2 to 100 millimeters) would have the diameter of really large trees. Imagine how each opens up soil particles as they go about their daily activities.

Finally, electrical charges on the surfaces of organic matter and clay particles attract each other in addition to chemicals (calcium, iron, aluminum) in water solution, acting as bonding agents that hold together soil particles.

Why are we going over this soil structure stuff? Because soil structure is a key characteristic of good growing conditions. If there is adequate soil structure, there is ample drainage between aggregates, but also plenty of plant-available capillary water. The air circulation necessary for biological activity is sufficient. And, perhaps most important, if there is adequate soil structure, there is space for soil biology to live. Good soil structure withstands torrential rains, the drying of desert-like droughts, herds of animal traffic, and deep freezes. Water and nutrient retention is high. Life in and on it thrives.

Poor soil structure results in a lack of water retention, and soil collapses under all the abovementioned environmental and man-made pressures. Little life is in it, and the serious reduction in fertility drives people to resort to chemical fertilizers in increasing amounts.

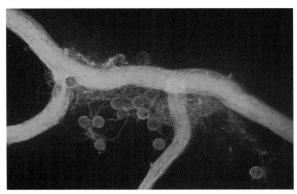

Microscopic view of a fungus growing on a corn root. The round bodies are fungal spores, the threads are fungal hyphae, and the green color is from dye-tagged glomalin, the glue that holds soil particles together. Photograph by Sara Wright, USDA-ARS.

Cation exchange capacity

All tiny particles, not just humus, carry electrical charges. These particles are called ions. Ions with a positive ($^+$) charge are called cations and negatively charged ($^-$) ones, anions. Positively charged particles are electrically attached to negatively charged particles. This is exactly what happens when opposite ends of magnets attract each other. When a positively charged cation attaches itself to a negatively charged anion, the cation is "absorbed" by the anion. Even microorganisms in the soil are small enough to carry and be influenced by electrical charges.

Sand particles are too large to carry electrical charges, but both clay and humus particles are small enough to have lots of negatively charged anions that attract positively charged cations. The cations that are absorbed by clay and humus include calcium (Ca^{++}), potassium (K^+), sodium (Na^+), magnesium (Mg^{++}), iron (Fe^+), ammonium (NH_4^+), and hydrogen (H^+). These are all major plant nutrients, and they are held in the soil by two components of good soil. The attraction of these cations to the clay and humus particles is so strong that when a solution containing them comes into contact, the attraction is satiated and only about 1% of the cation nutrients remains in solution.

There are anions in soil as well. These include chloride (Cl^-), nitrate (NO_3^-), sulfate (SO_4^-), and phosphate (PO_4^-)—all plant nutrients. Unfortunately, soil anions are repelled by the negative charge on clay and humus particles and therefore stay in solution instead of being absorbed. These plant nutrients are often missing from garden soils, as they are easily leached away in the soil solution when it rains or soil is watered: nothing is holding them on to soil surfaces.

Why does this matter? The surfaces of root hairs have their own electrical charges. When a root hair enters the soil, it can exchange its own cations for those attached to clay or humus particles and then absorb the cation nutrient

SOIL TEXTURE	CEC (MEQ/100G)
Sands (light colored)	3–5
Sands (dark colored)	10–20
Loams	10–15
Silt loams	15–25
Clay and clay loams	20–50
Organic soils	50–100

Cation exchange capacities for various soil textures. Courtesy Tom Hoffman Graphic Design.

involved. Roots use hydrogen cations (H^+) as their exchange currency, giving up one hydrogen cation for every cation nutrient absorbed. This keeps the balance of charges equal. This is how plants "eat."

The place where the exchange of a cation occurs is known as a cation exchange site, and the number of these exchange sites measures the capacity of the soil to hold nutrients, or the cation exchange capacity (CEC). A soil's CEC is simply the sum of positively charged nutrient replacements that it can absorb per unit weight or volume. CEC is measured in milligram equivalents per 100 grams (meq/100g). What the gardener needs to know is that the higher the CEC number, the more nutrients a soil can hold and therefore, the better it is for growing plants. The higher the CEC, the more fertile the soil. You can order a CEC test to be run by a professional soil lab.

The CEC of soil depends, in part, on its texture. Sand and silt have low CECs because these particles are too big to be influenced by an electrical charge and hold nutrients. Clay and organic particles impart a high CEC to soils because they do carry lots of electrical charges: the more humus and, to a point, clay present in soils, the more nutrients can be stored in the soil, which is why gardeners seek more organics in their soils.

There are limits to a good thing. Don't forget that clay particles are extremely small; too much clay and too little humus results in a high CEC but little air in the soil, because the pore space is too small and cut off by the clay's platy structure. Such soil has good CEC but poor drainage. Thus it is not enough to know the CEC alone; you have to know the soil texture and mixture.

Soil pH

Most of us have a basic understanding of pH as a way to measure liquids to see if they are acid or not. On a scale of 1 to 14, a pH of 1 is very acidic and a pH of 14 is very alkaline (or basic), the opposite of acidic. The pH tells the concentration of hydrogen ions (H^+, a cation) in the solution being measured. If you have a lot of hydrogen ions compared to the rest of what is in solution, the pH is low and the solution is acidic. Similarly, if you have relatively few hydrogen ions in solution, then you have a solution with a high pH, one that is alkaline.

As a gardener, you (fortunately) don't need to know much more about pH. You do need to understand, however, that every time a plant root tip exchanges a hydrogen cation for a nutrient cation, the concentration of hydrogen ions in the solution increases. As the concentration of H^+ goes up, the pH goes down— the soil is increasingly acidic. Things usually balance out, however, because root surfaces also take up negatively charged anions, using hydroxy (OH^-)

anions as the medium of exchange. Adding OH^- to the solution raises the pH (that is, soil is increasingly alkaline) because it lowers the concentration of H^+. Fungi and bacteria are small enough to have cations and anions on their surfaces, electrically holding or releasing the mineral nutrients they take in from decomposition in the soil. This, too, has an impact on the pH of the soil.

Why is pH a consideration when we talk about the soil food web? The pH created by nutrient-ion exchanges influences what types of microorganisms live in the soil. This can either encourage or discourage nitrification and other biological activities that affect how plants grow. As important, each plant has an optimum soil pH. As you will learn, this has more to do with the need of certain fungi and bacteria important to those plants to thrive in a certain pH than it does with the chemistry of pH.

Knowing your soil's pH is useful in determining what you want to put into your soil, if anything, to support specific types of soil food webs. And knowing the pH in the rhizosphere helps determine if any adjustments should be made to help plant growth.

The rest of Part 1 covers the biology that lives in the soil. You have to appreciate the soil first, however.

Chapter 3

Bacteria

BACTERIA are everywhere. Few gardeners appreciate that they are crucial to the lives of plants, and fewer still have ever taken them into consideration. Yet no other organism has more members in the soil, not even close. In part, this is because these single-celled organisms are so minuscule that anywhere from 250,000 to 500,000 of them can fit inside the period at the end of this sentence.

Bacteria were the earliest form of life on earth, appearing at least 3 billion years ago. They are prokaryotes: their DNA is contained in a single chromosome that is not enclosed in a nucleus. Their size, or more precisely their lack thereof, must be the main reason our familiarity with bacteria is usually limited to the diseases they cause and the need to wash our hands before eating. Most baby boomers used a standard-issue 1000 power microscope to study microorganisms, but bacteria are too small to see in any detail at this power. School microscopes have gotten better, and some lucky students now do get a closer look, literally, at bacteria. The three basic shapes, all represented in the soil, are coccus (spherical or oval), bacillus (rod-shaped), and spiral.

Bacteria reproduce, for the most part, by single cell division; that is, one cell divides and makes two cells, they each divide again, and so forth. Amazingly, under laboratory conditions, one solitary bacterium can produce in the vicinity of 5 billion offspring in a mere 12 hours if they have enough food. If all bacteria reproduced at this rate all the time, it would take only a month or so to double the mass of our planet. Fortunately, soil bacteria are limited by natural conditions, predators (protozoa chief among them), and a slower reproductive rate than their laboratory cousins; for example, bacteria must have some form of moisture for the uptake of nutrients and the release of waste. In most cases, moisture is also required for bacteria to move about and to transport the enzymes they use to break down organic matter. When soils become too dry, many soil bacteria go dormant. Bacteria, incidentally, rarely die of old age, but are usually eaten by something else or killed by environmental changes and then consumed by other decomposers, often other bacteria.

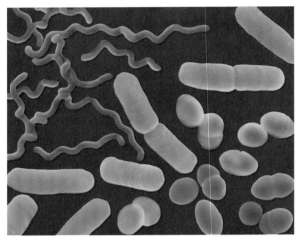

Photo composite of the three basic shapes of bacteria: coccus, bacillus, and spiral, 800×. Image copyright Dennis Kunkel Microscopy, Inc.

Primary decomposers

Despite their tiny size, bacteria are among the earth's primary decomposers of organic matter, second only to fungi. Without them, we would be smothered in our own wastes in a matter of months. Bacteria decompose plant and animal material in order to ingest nitrogen, carbon compounds, and other nutrients. These nutrients are then held immobilized inside the bacteria; they are released (mineralized) only when the bacteria are consumed or otherwise die and are themselves decayed.

Different kinds of soil bacteria survive on different food sources, depending on what is available and where they are located. Most, however, do best decomposing young, still-fresh plant material, which composters call green material. Green material contains lots of sugars, which are easier for bacteria to digest than the more complex carbon compounds of other plant material. Composters call this brown material, and until it is broken down into smaller carbon chains, other members of the soil food web more readily digest this than do bacteria.

Given their diminutive size, bacteria must ingest what are necessarily even more tiny pieces of organic matter. How do they do this? The short answer is they take in food directly through their cell walls, which are composed, in part, of proteins that assist in this molecular transport. On the inside of a bacterium's cell wall is a mixture of sugars, proteins, carbons, and ions—a rich soup that is out of equilibrium with the less concentrated mixture outside the cell wall. Nature likes to try to keep things balanced; normally, water would flow from the dilute solution without into the more concentrated one within (a special form of diffusion known as osmosis), but in the case of bacteria, cell walls act as osmotic barriers.

Molecular transport across the cellular membrane is accomplished in several ways. In the most important, active transport, the membrane proteins act as molecular pumps and use energy to suck or push their target through the cell wall—nutrients in, waste products out. Different proteins in the membrane transport different kinds of nutrient molecules. One way to imagine this is to think of an old-fashioned fire bucket brigade, in which the water was passed from its source to the fire: these proteins pass "buckets" of nutrients into the cell.

Active transport is a fascinating but complicated process fueled by electrons located on both sides of the membrane surface. The gardener should certainly be aware of and appreciate how bacteria feed but only needs to understand that bacteria break up organic matter into small, electrically charged pieces and then transport these through their cellular membranes, ready for use. Once inside the bacteria, the nutrients are locked up.

Other members of the soil food web obtain their energy and nutrients by eating bacteria. If there aren't sufficient numbers of bacteria in the soil, populations of these members of the soil food web suffer. Bacteria are part of the base of the soil food web food pyramid.

Feeding bacteria

Root exudates are favorite foods for certain soil bacteria, and as a result, huge populations of them concentrate in the rhizosphere, where bacteria also find nutrition from the cells sloughed off during root-tip growth. But not all soil bacteria live in the rhizosphere, for, fortunately, organic matter is almost as ubiquitous as bacteria. All organic matter is made up of large, complex molecules, many of which consist of chains of smaller molecules in repetitive patterns that usually contain carbon. Bacteria are able to break the bonds along certain points of these chains, creating smaller chains of simple sugars and fatty and amino acids. These three groups provide the basic building blocks bacteria need to sustain themselves.

Bacteria use enzymes both to break the bonds holding organic chains together and to digest their food. All this is done outside the organism before ingestion. Untold numbers of enzymes are employed by bacteria, who have adapted over the millenia to attack all manner of organic and even inorganic matter. It is an astonishing feat that bacteria can employ enzymes to break down organic matter, while at the same time not impacting their own cell membrane.

Air and no air

There are two main groups of bacteria. The first, anaerobic bacteria, are able to live in the absence of oxygen; indeed, most cannot live in its presence. The bacterial genus *Clostridium*, for example, does not need oxygen to survive and can invade and destroy the inside soft tissue of decaying matter. By-products of anaerobic decay include hydrogen sulfide (think rotten eggs), butyric acid (think vomit), ammonia, and vinegar. The notorious *Escherichia coli* (*E. coli*) and other bacteria normally found in the mammalian gastrointestinal tract (and thus in poorly made, manure-based composts) are facultative anaerobes, meaning they can live in aerobic conditions if they must but prefer anaerobic environments.

Most gardeners have smelled by-products of anaerobic decomposition, perhaps in the garden but certainly in the refrigerator. These are smells to remember when composting and gardening with the soil food web because anaerobic conditions foster pathogenic bacteria and, worse, kill off beneficial aerobic bacteria, the other major group of bacteria: those that require air.

While some facultative aerobic bacteria are able to live in anaerobic conditions if they must, most cannot. Aerobic bacteria are not normally known to cause bad smells. In fact, the actinomycetes (of order Actinomycetales, specifically the bacterial genus *Streptomyces*) produce enzymes that include volatile chemicals that give soil its clean, fresh, earthy aroma. Anyone who has gardened recognizes this smell, the smell of "good soil."

Actinomycetes are different from other soil bacteria: they actually grow filaments, almost like fungal hyphae. Some scientists believe *Streptomyces* species use their branching filaments to connect soil particles so they, along with the

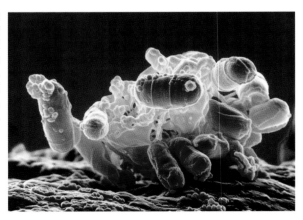

Low-temperature electron micrograph of a cluster of *E. coli* bacteria. Individual bacteria in this photo are oblong and colored brown. Photograph by Eric Erbe, digital color by Christopher Pooley, USDA-ARS.

soil particles, become too big to be eaten by their natural predators, the proto-zoan ciliates, which would engulf and ingest them. Actinomycetes are particu-larly adept at decaying cellulose and chitin—two difficult-to-digest ("brown") carbon compounds, the former found in plant wall cells and the latter in fun-gal cell walls and in arthropod shells. These are not normal foods of other bac-teria. Actinomycetes are also adapted to live in a wider range of pH than other bacteria, from acidic to alkaline.

Decay of cellulose

Cellulose, a complex carbohydrate made up of long chains of carbon-based glucose, is the molecular material that gives plants structure. It constitutes half the mass of plant bodies, and hence half the mass of organic matter created by plants. Specialized bacteria, like the aptly named *Cellulomonas*, carry cellulose-breaking enzymes that they release only when they come into contact with cel-lulose, as opposed to the random release of enzymes by other bacteria who eat in a hit-and-miss manner.

Most bacteria reach their limit when it comes to the noncarbohydrate lignin, another prevalent, molecularly complex plant material. Lignin, the tough brown component of barks and woody materials, is a much more com-plex organic molecule than cellulose, made up of chains of interlinked alco-hols; these are resistant to the enzymes produced by most bacteria and are left for fungi to decay.

Element cycling

One way of looking at decay is to view it as nature's recycling system. Bacteria in the soil food web play a crucial role in recycling three of the basic elements needed for life: carbon, sulfur, and nitrogen. For example, CO_2 (carbon diox-ide) is a major by-product of aerobic bacterial metabolism. Carbon tied up in plant and animal biomass is cycled into CO_2 gas during decay. Photosynthesis in higher plants converts the CO_2 into organic compounds, which are eventu-ally consumed and then recycled back to CO_2.

Similarly, sulfur is recycled. Sulfur-oxidizing bacteria use the element to make plant-available, water-soluble sulfates. Liberated from organic materials by anaerobic bacteria, sulfur-containing compounds are produced by chemo-autotrophs, bacteria that get energy from the oxidation of sulfur.

The nitrogen cycle, propelled in part by specialized bacteria, is one of the most important systems in the maintenance of terrestrial life: living organisms produce the vital organic compounds, the building blocks of life—amino and nucleic acids—using nitrogen. The strong bonds holding atmospheric nitrogen molecules together make this nitrogen inert for all practicable purposes and useless for plant needs. For plants to be able to use nitrogen, it has to be "fixed"—combined with either oxygen or hydrogen—producing ammonium (NH_4^+), nitrate (NO_3^-), or nitrite (NO_2^-) ions. This important process is called nitrogen fixation.

Certain bacteria convert nitrogen from the atmosphere into plant-available forms. The genera that accomplish this nitrogen-fixing feat are *Azotobacter*, *Azospirillum*, *Clostridium*, and *Rhizobium* (any one of which would be a great name for a comic book superhero). *Azotobacter*, *Azospirillum*, and *Clostridium* live free in the soil; *Rhizobium* species actually live in the root tissues of certain plants, particularly legumes, where they form visible nodules.

We don't mean to suggest you need to memorize the species of soil bacteria, but we *do* want you to focus on the fact that nitrogen fixation as well as the recycling of carbon and sulfur requires the interventions of living organisms. These are always taught as chemical processes, but they are really biological.

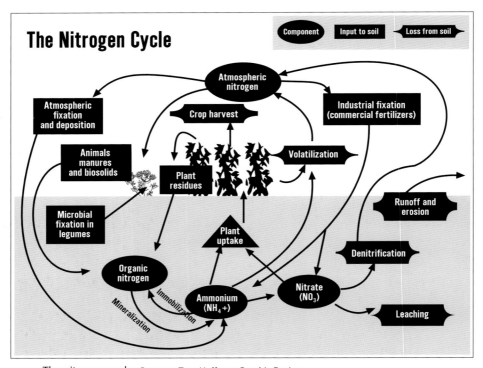

The nitrogen cycle. Courtesy Tom Hoffman Graphic Design.

Bacteria carry out these processes in the soil, forming symbiotic relationships with specific plants or existing symbiotically within organisms. Sounds like a case of the biology creating the chemistry to us.

Another part of the nitrogen cycle, the place at which it "starts" in the soil, involves the decomposition of proteins into ammonium (NH_4^+). This ammonium usually figures as part of the waste product produced by protozoa and nematodes after eating bacteria and fungi. Next, special nitrite bacteria (*Nitrosomonas* spp.) convert the ammonium compounds into nitrites (NO_2). A second type of bacteria, nitrate bacteria (*Nitrobacter* spp.), convert the nitrites into nitrates (NO_3^-).

Nitrifying bacteria do not generally like acidic environments; their numbers (and hence the conversion of nitrogen into nitrates) therefore diminish when soil pH drops below 7. Bacterial slime (already mentioned for its ability to bind soil particles together) happens to have a pH above 7. Thus, if there are enough bacteria in an area, the slime they produce keeps the pH in their vicinity above 7, and nitrification can occur. If not, the ammonium first produced by organisms in the soil is not all converted to nitrate form. If the pH is 5 or lower, very little if any of the ammonium is converted.

Denitrifying bacteria convert nitrogen salts back to N_2, which escapes into the atmosphere. Obviously, denitrifying bacteria do not help the fertility of soil, but they are essential in that they keep the nitrogen cycle moving.

Biofilms

Bacterial slime, or biofilm, is a matrix of sugars, proteins, and DNA. The fact that bacterial slime in the soil is slightly alkaline not only influences the pH where it counts most, in the rhizosphere, but also buffers the soil in the area, so the pH remains relatively constant.

Some bacteria use their film as a means of transportation, literally squirting this substance as a means of propulsion. (Most bacteria, however, travel using an astonishing bit of natural nanotechnology—with the aid of one or more whip structures, or flagella, that resemble and operate like propellers.) Biofilms save bacteria from desiccation as the soil dries; soil bacteria often live inside sticky globs of biofilms, complete with an infrastructure of channels filled with water for transport of nutrients and wastes. Biofilms can also be a defense against antibiotics produced by other organisms, including fellow bacteria. Bacteria colonies protected by slime are 1000 times more resistant than individual bacteria to antibiotics and microbicides.

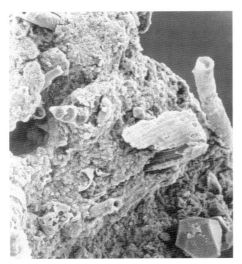

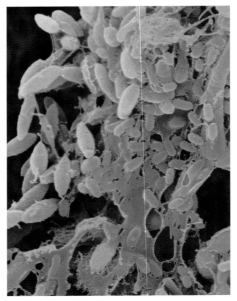

Scanning electron micrograph (SEM) of a biofilm surface. Insect parts and plant fibers are embedded in the slime, along with numerous crystals. Photograph by Ralph Robinson, www.microbelibrary.org.

Bacterial biofilm on stainless steel, 1600×. Image copyright Dennis Kunkel Microscopy, Inc.

Nutrient retention

Bacteria play a major role in plant nutrition. They lock up nutrients that might otherwise disappear as a result of leaching. They do so by ingesting them while decomposing organic matter and retaining them in their cellular structures. Since the bacteria are themselves attached to soil particles, the nutrients remain in the soil instead of being washed away, as is the case with chemical fertilizers.

Indeed, these nutrients will be tied up, immobilized inside the bacteria until the bacteria are eaten and reduced to wastes. Since soil bacteria don't travel very far, and there is ample source of bacterial food in the root zone, the nutrients ingested by the bacteria are kept in the vicinity of the roots. Other organisms, such as protozoa, play major roles consuming bacteria, releasing excess nitrogen as ammonium (NH_4^+) in their wastes, which are deposited in the rhizosphere, right where the roots can absorb nutrients.

Other benefits of soil bacteria

Some anaerobic bacteria produce alcohols that are toxic to plant life and to other bacteria. These anaerobic bacteria can be avoided when gardening by controlling the conditions that allow them to multiply: poor soil texture, lack of pore space, standing water, and compacted soil. Other bacteria are pathogens that cause disease in higher plants. The list of pathogenic bacteria is a long one, including bacteria that cause citrus canker, diseases of potatoes, melons, and cucumbers, and fire-blight of pears, apples, and the like. Thousands of bacterial pathogens are in soil, and billions of dollars are spent every year to protect crops from damage by the culprit bacteria. *Agrobacterium tumefaciens* causes galls or tumors to grow on the stems of certain plants. *Burkholderia cepecia* is a bacterium that infects and rots the roots of onions. Some *Pseudomonas* species cause leaf curl and black spot on tomatoes.

Despite the presence of pathogenic bacteria, there are more benefits to a healthy soil bacteria population than not. For example, bacterial activity is also often responsible for breaking down pollutants and toxins. These processes are usually aerobic, requiring oxygen to occur. You undoubtedly have heard of bacteria that can eat oil spilled on a beach in Alaska; there are similar bacteria that will eat gasoline spilled on your lawn, for example.

Soil bacteria produce many of the medicinal antibiotics upon which we have come to depend. One can only speculate that since these bacteria have to compete not only with other bacteria for nutrients but also with fungi and other organisms, they evolved protective capabilities. For example, *Pseudomonas* bacteria can correct take-all, a disastrous fungal wheat disease, by producing phenazines, very strong, broad-spectrum antibiotics. Obviously, many soil bacteria keep pathogenic bacteria in check, a big benefit of a healthy soil food web.

All bacteria compete with each other and with other organisms for the finite amount of food the soil offers and thus keep each other's populations in balance. Soils with a high diversity of bacterial types are more likely to have a larger number of nonpathogenic bacteria outcompeting pathogenic bacteria for space and nutrients. We are convinced that using the soil food web's natural defenses is the best way to keep the bad guys in check. Gardeners need to appreciate that bacteria are at the front line of defense.

Chapter 4

Fungi

OVER 100,000 different kinds of fungi are known, and some authorities suggest a million more are out there waiting to be discovered. Say the word, however, and most gardeners immediately think of the familiar white toadstools, bracket and coral fungi, and puffballs that appear in the lawn or on the bark of trees (or they know soil fungi from the diseases they cause—more on this, later in the chapter). But except for the white threads and the spore-producing mushrooms, soil fungi are as invisible as bacteria, requiring a microscope of several hundred power to be seen. Even the visible congregations of mycelia are usually hidden in the organic matter they are in the process of decaying.

Fungi, too, are underappreciated by gardeners, and yet they play a key role in the soil food web and are an important tool for those who garden using soil food web principles. It wasn't too long ago that they were considered plants without chlorophyll and included for classification purposes in the plant kingdom; but because fungi are unable to photosynthesize, and build their cell walls from chitin instead of cellulose, among other unique characteristics, they are now placed in their own kingdom in the domain Eukarya.

Fungi, like higher plants and animals, are eukaryotes: organisms that have cells with distinct, enclosed nuclei. Each cell can have more than one nucleus. Fungi usually grow from spores into thread-like structures called hyphae (singular, hypha). A single hyphal strand is divided into cells by walls, or septa (singular, septum). The walls connecting hyphae cells are seldom completely sealed off from other cells in the strand, thus allowing liquids to flow between cells. Masses of invisible hyphae growing in close enough proximity form visible threads, or mycelia (singular, mycelium), which you may have seen in decomposing leaf litter. Fungi reproduce in many different ways, not just by spores, but never by seeds, as the most advanced plants commonly do.

A fungal hypha is considerably larger than a bacterium, the average length being 2 to 15 micrometers with a diameter of 0.2 to 3.5 micrometers—still so thin that it takes hundreds of thousands of individual hyphal strands to form a network thick enough for the human eye to see. A teaspoon of good garden soil may contain several yards of fungal hyphae, invisible to the naked eye; millions upon millions merge together to produce something as obvious as a king bolete or an intricate *Amanita muscaria* in all their fruiting glory. These and

Amanita muscaria, the beautiful but poisonous fly agaric. Photograph by Judith Hoersting.

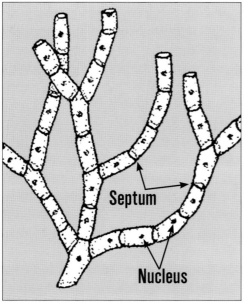

A diagram of a hypha. Courtesy Tom Hoffman Graphic Design.

other mushrooms are simply the fruiting bodies of fungi. Consider the energy and nutrients required to produce them.

One major advantage fungi have over bacteria, and perhaps the reason they were misclassified for so long as plants, is the ability of fungal hyphae to grow in length. Unlike bacterial cells, whose world is a very finite one, fungal hyphae can travel over space measured in feet or meters, distances that for a bacterium are truly epic. And unlike bacteria, fungi do not need a film of water in order to spread through the soil. Fungal hyphae are thus able to bridge gaps and go short distances, which allows them to locate new food sources and transport nutrients from one location to another, relatively far away from its origin.

The ability to transport nutrients is another key difference between fungi and bacteria. Fungal hyphae contain cytoplasm, a liquid circulated throughout the septa in their cells. When a hyphal tip invades a nematode, for example, it drains its hapless victim of its nutrients and distributes them in the hyphal cytoplasm and from there through the main body of the fungus. Nutrients are thus transferred from the tip of the fungal hypha to a wholly new location that can be several yards away (think conveyor belt). Once inside the fungus, the nutrients are immobilized and will not be lost from the soil.

Fungi produce special structures—for example, mushrooms above ground or truffles below—to disperse spores. Since fungi grow in all sorts of environments, they have devised some elaborate methods to achieve spore dispersal, including attractive scents, triggers, springs, and jet propulsion systems. To ensure survival, fungal spores can develop tough membranes that allow them to go dormant for years if the conditions are not right for immediate germination.

As with bacteria, fungi occur universally; some species even exist in the

Fungal spores are produced in bodies that rise above the fungus to help dispersal. Courtesy T. Volk. Reprinted, with permission, from http://www.apsnet.org/, American Phytopathological Society, St. Paul, Minnesota.

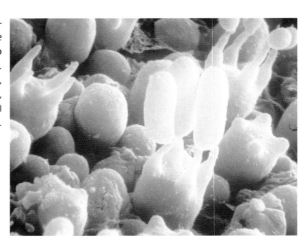

frozen region of Antarctica. Airborne dispersal of spores helps explain why visitors from, say, Alaska, will recognize species of fungi growing in far-off Australia. While dormant spores can be found around the world, they need the right conditions to germinate and grow. Thus, fungal spores may be found continents away from their source, but they may not be functional because the conditions for growth are not right.

Fungal growth and decay

While some fungi prefer the "softer," easier-to-digest sugars characteristic of the foods that feed bacteria, most go for tougher-to-digest foods (mainly because bacteria are better and faster at grabbing and taking up the simple sugars). Fungi, however, win in the competition for more complex foods: they produce phenol oxidase, a strong enzyme that dissolves even lignin, the woody compound that binds and protects cellulose. Another characteristic of fungi is their ability to penetrate hard surfaces. Fungi have perfected apical growth—that is, growth at their hyphal tip. Apical or tip growth is an incredibly complex process, an engineering job akin to building a tunnel under a river and requiring great coordination of events. Even before electron microscopes, scientists identified a dark spot, the Spitzenkörper, in the tip of actively growing hyphae; when hyphal growth stopped, the Spitzenkörper disappeared. It seems this mysterious region has something to do with controlling or perhaps directing apical growth.

During apical growth, new cells are constantly being pushed into the tip and along the sidewalls, elongating the hyphal tube. Materials for the growth of fungal hyphae are supplied to the advancing tip by the cytoplasm, which transports vesicles loaded with all necessary "construction" supplies. Of course, it is important to keep extraneous material from flowing into the hypha as well as out while this growth is happening. All the while, powerful enzymes capable of dissolving all but the most recalcitrant carbon compounds are released as the new cells are put into place. Think about it: these enzymes are powerful enough to convert lignin, cellulose, and other tough organic matter into simple sugars and amino acids, yet they do not decay the chitin cell walls of the fungi.

Fungi can grow up to 40 micrometers a minute. Discount for the moment the speed, which is incredibly fast for such tiny organisms, and compare the distance covered to the movement of a typical soil bacterium, which may travel only 6 micrometers in its entire life.

As with the death of any organism in the soil, the death of fungi means the

nutrients contained within them become available to other members of the soil food web. But when fungi die, their hyphae leave a subway system of microscopic tunnels, up to 10 micrometers in diameter, through which air and water can flow. These "tubes" are also important safety zones for bacteria trying to elude protozoa: protozoa are considerably bigger than the tunnels.

Fungi are *the* primary decay agents in the soil food web. The enzymes they release allow fungi to penetrate not only the lignin and cellulose in plants (dead or alive) but also the hard, chitin shells of insects, the bones of animals, and—as many gardeners have learned—even the protein of strong toenails and fingernails. Bacteria can hold their own, but they require simpler-to-digest foods, often the by-products of fungal decay, and often only after such food has been broken or opened up by fungi and others. Compared to fungi, bacteria are in the Minor Leagues of decaying ability.

Fungal feeding

The acidic digestive substances produced by fungi and leaked out of their hyphal tips are similar to those utilized by humans; fungi don't require a stomach as a vessel in which to digest food, however. Like bacteria, fungi lack mouthparts; instead, fungal decay breaks up organic materials into compounds the fungus can then ingest through its cell walls via diffusion (osmosis) and active transport. Nutrients taken in by fungi are usually immobilized, just as they are when ingested by bacteria, and later released. Like bacteria, then, fungi should be viewed as living containers of fertilizer.

Excess acids, enzymes, and wastes are left behind as the fungus continues to grow and as a consequence, the digestion of organics continues even though the fungus has moved on, opening up organic material for bacterial decay and making nutrients available to plants and others in the soil community. Hyphal growth gives a fungus the ability to move relatively long distances to food sources instead of waiting for its food to come close (though it can clearly do this, too, as the nematode-trapping fungus proves). Fungi can, for example, extend up into the leaf litter on the surface of the soil, decay leaves, and then bring the nutrients back down into the root zone—a huge advantage over bacteria, the other primary nutrient recycler in the soil food web.

Soil fungi are usually branched and quite capable of gathering organic compounds from different sources simultaneously. Once the nutrient material is inside the cellular membrane, it is transported back through the network of fungal hyphae that often ends at the root of a plant, where some fungi trade for exudates. Thus the same fungus can extend hyphae downward and outward, absorbing several crucial nutrients—phosphorus, copper, zinc, iron, nitrogen—as well as water. In the case of phosphorus, for example, the propensity of fungi to gather and transport it over distances is truly remarkable. This mineral is almost always chemically locked up in soils; even when it is applied as fertilizer, phosphorus becomes unavailable to plants within seconds. Not only do fungi seek out this necessary plant nutrient, but they have the ability to free it from its chemical and physical bonds. Then they transport their quarry back to plant roots, where the phosphorus is absorbed and utilized.

Don't forget that in those instances where a fungus brings food back to a plant root tip, it was attracted to that plant by the plant's exudates. Fungi are good, but the plant is in control.

Fungi and plant-available nitrogen

Some fungi trade nutrients for exudates, but most often nutrients are released as waste after they are consumed by fungi or when the fungi die and are decayed. Much of what is released is nitrogen. A key tenet of gardening with the soil food web is that plants can take up nitrogen in two forms, either as ammonium ions (NH_4^+) or as nitrate ions (NO_3^-). The nitrogen released by fungi is in ammonium form (NH_4^+). If nitrifying bacteria are present, this is converted in two steps to nitrate (NO_3^-).

The enzymes produced by fungi are decidedly acidic and lower the pH. Remember that bacterial slime raises soil pH; nitrogen-fixing bacteria generally require a pH above 7. As soils become dominated by fungi, the populations of nitrogen-fixing bacteria required to convert ammonium into nitrates diminish because the pH is lowered by the acids the fungi produce. More ammonium therefore remains as plant-available ammonium instead of being converted to nitrates. This has an important implication to gardening with the soil food web: fungally dominated soils tend to have nitrogen in ammonium form. This is great if you are a plant that prefers ammonium to nitrate, but not so good if you prefer to have your ammonium converted to nitrates (who wants what is explained in chapter 12).

Fungal adaptations

Fungi have developed all sorts of clever strategies to make it through life—our nematode-strangling fungus proves it. The fungus that developed this very artful and useful adaptation is *Arthrobotrys dactyloides*. The ring that trapped the nematode is actually just a hyphal branch, twisted back on itself. These branches each consist of only three cells, which, when touched, produce a signal to let water in; the cells then swell to three times their size and the unsuspecting victim is killed in a tenth of a second. Pretty amazing—a sophisticated trapping mechanism developed from an inverted branch using only three cells. Once again, nanotechnology can only hope to duplicate such a complicated process. Not only does the fungus figure out a way to kill nematodes, which are all blind, but it attracts them to its trap in the first instance. In this case, the fungus releases a chemical that attracts the worm.

Within a matter of only a few minutes after trapping, the tip of a fungal hypha enters the nematode's body, secretes its powerful enzymes, and starts absorbing nutrients. As this is exactly what the nematode has been doing—eating—the worm is usually a real treasure trove of nutrients for the fungus. These nutrients, of course, are then locked up inside the fungus until the fungus is eaten by one of its predators or it trades them for exudates. Then the nutrients are mineralized and again are available to plants.

The fungus *Pleurotus ostreatus*, the common oyster mushroom you can buy at the supermarket, uses another clever technique to trap food. It emits toxic drops from the tips of its hyphae; an unsuspecting nematode (our perennial fungal fall guy), out and about, looking for food, touches a drop with its mouth and within minutes is immobilized. A few hours later, and the fungus is inside the nematode, already digesting it.

This is not a bad way to ensure a meal: attract your food and either trap it or stun it and then consume it. Other mechanisms have evolved as well. Some fungi use adhesives to stick to nematodes. Other soil fungi trap protozoa and even springtails, much larger microarthropods that are big enough to see with the naked eye. Once attached, the fungi digest their prey and again lock up or immobilize plant nutrients.

What drives soil fungi in the direction of particular nutrients is still an open question. It is known that some send out filaments as if they were scouts looking for nutrients. If you have ever seen a well-trained bird dog look for a downed bird, you get the idea. The dog circles until its nose finds the bird. Some fungi clearly possess tactile- or contact-sensing capabilities that allow them to orient in a certain direction so they can invade their prey or other food source. Others demonstrate the ability to track specific chemicals they know to be in the vicinity of specific prey.

For the gardener it is sufficient to know that fungi can find nutrients. When a source is found, fungal strands head over to the area and literally settle in, digesting the material, often combining one nutrient source material with another and transporting nutrients back to the base of the fungus. All the while, other strands "scout" for more food to attack. Nutrients are held inside cell walls, preventing them from leaching away.

Fungi and symbiosis

Soil fungi also form two extremely important mutual relationships with plants. The first is the association of certain fungi with green algae, which results in the formation of lichens. In this symbiotic relationship, the fungus gets

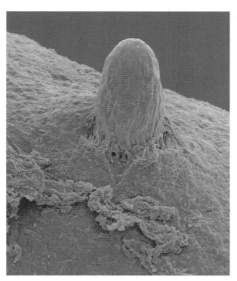

A small branch protruding from the main thallus of a tree lichen, 140×.
Image copyright Dennis Kunkel Microscopy, Inc.

food from the alga, which utilizes its photosynthetic powers while the fungal strands make up the thallus, or body, of the lichen, in which the pair lives. Chemicals secreted by the fungus break down the rock and wood upon which the lichens grow. This creates minerals and nutrients for soil, soil microbes, and plants.

The second are mycorrhizae (from the Greek for "fungus-root"), symbiotic associations between plant roots and fungi. In return for exudates from plant roots, mycorrhizal fungi seek out water and nutrients and then bring them back to the plant. The plant becomes dependent on the fungi, and the fungi, in turn, cannot live without the plant's exudates. It is a wonderful world, indeed.

Mycorrhizae have been known since 1885, when German scientist Albert Bernhard Frank compared pines grown in sterilized soil to those grown in sterilized soil inoculated with forest fungi. The seedlings in the inoculated soil grew faster and much larger than those in the sterilized soil. Yet it was only in the 1990s that the terms mycorrhiza (the symbiotic root-fungus relationship; plural, mycorrhizae) and mycorrhizal (its associated adjective) started to creep into the agricultural industry's lexicon, much less the home gardener's.

We're the first to admit that we were blindsided by the subject—and one of us had written a popular garden column every week for 30 years and never once mentioned them out of sheer ignorance, a state shared with most gardeners. We now know the extent of our ignorance: at least 90% of all plants form mycorrhizae, and the percentage is probably 95% and even higher. What is worse, we learned that these relationships began some 450 million years ago, with terrestrial plant evolution: plants started growing on the earth's surface only after fungi entered into relationships with aquatic plants. Without mycorrhizal fungi, plants do not obtain the quantities and kinds of nutrients needed to perform at their best; we must alter our gardening practices so as not to kill these crucial beneficial fungi.

Perhaps gardeners lack appreciation for fungi because all soil fungi are very fragile. Too much compaction of soil, and fungal tubes are crushed and the fungi killed. Clearly fungicides but also pesticides, inorganic fertilizer, and physical alteration of the soil (rototilling, double digging) destroy fungal hyphae. Chemicals do so by sucking the cytoplasm out of the fungal body. Rototilling simply breaks up the hyphae. The fruiting bodies of mycorrhizal fungi even decrease when fungi are exposed to air pollution, particularly that containing nitrogenous substances.

Mycorrhizal fungi are of two kinds. The first, ectomycorrhizal fungi, grow close to the surface of roots and can form webs around them. Ectomycorrhizal fungi associate with hardwoods and conifers. The second are endomycorrhizal

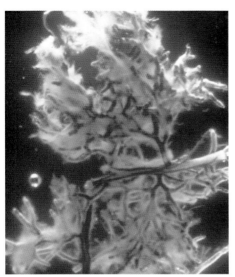

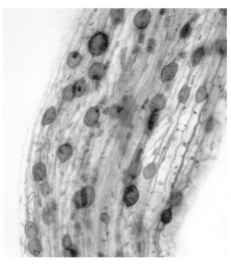

Endomycorrhizal fungi penetrating roots. Courtesy L. H. Rhodes. Reprinted, with permission, from http://www.apsnet.org/, American Phytopathological Society, St. Paul, Minnesota.

Ectomycorrhizal fungi forming a dense white net around roots. Courtesy Mycorrhizal Applications, www.mycorrhizae.com.

fungi. These actually penetrate and grow inside roots as well as extend outward into the soil. Endomycorrhizal fungi are preferred by most vegetables, annuals, grasses, shrubs, perennials, and softwood trees.

Both types of mycorrhizal fungi can extend the reach as well as the surface area of plant roots; the effective surface area of a tree's roots, for example, can be increased a fantastic 700 to 1000 times by the association. Mycorrhizal fungi get the carbohydrates they need from the host plant's exudates and use that energy to extend out into the soil, pumping moisture and mining nutrients from places the plant roots alone could not access. These fungi are not lone miners, either. They form intricate webs and sometimes carry water and nutrients to the roots of different plants, not only the one from which they started. It is strange to think of a mycorrhizal fungus in association with one plant helping others at the same time, but this occurs.

Finding and bringing back the phosphorus that is so critical to plants seems to be a major function of many mycorrhizal fungi; the acids produced by mycorrhizal fungi can unlock, retrieve, and transport chemically locked-up phosphorus back to the host plant. Mycorrhizal fungi also free up copper, calcium, magnesium, zinc, and iron for plant use. As always, any nutrient compounds not delivered to the plant roots are locked up in the fungi and are released when the fungi die and are decayed.

Pathogenic and parasitic fungi

Beneficial fungi compete for nutrients and form protective webs and nets, often in conjunction with bacteria, around roots (and even on leaf surfaces, as leaves produce exudates that attract bacteria and fungi as well); this prevents some of their pathogenic and parasitic fungal cousins from invading the plant. The list of fungal pathogens impacting agricultural and horticultural crops is long; the topic fills many books and is beyond the scope of this one. Smut fungi, for example, impact the flowers of cereal grains. Rust fungi cause diseases on wheat, oats, rye, fruits, and pines. More common garden problems are downy mildew (*Plasmopara* spp., *Sclerophthora* spp.), root rots (*Phytophthora* spp.), and white rusts (*Albugo* spp.).

Be there a gardener who has not encountered botrytis or powdery mildew, a catch-all name for a group of fungi that infects different plants with the same results, an unsightly gray or white powdery fungal growth that covers leaves, stems, and flowers? Most powdery mildew fungi produce airborne spores that

Gray mold fungus (*Botrytis cinerea*) attacking a strawberry plant. Photograph by Scott Bauer, USDA-ARS.

do not require free water to germinate. Given temperatures between 60 and 80F (15 and 27C) and high humidity, these spores germinate and infect their hosts in your yard. How about fusarium wilt on tomatoes, the first thing to suspect when a tomato's leaves start to yellow from the bottom of the plant up? It is caused by *Fusarium oxysporum* f. sp. *lycopersici*, a soil-borne fungus that can survive for a decade or more in dormant stages. It enters the plant through roots and invades its water distribution network. Further testament to the power of fungi is *Armillaria mellea* (oak root fungus), which causes oak death—a tiny fungus taking down towering oaks. The fungal activity decays a tree's lignin and cellulose to such an extent that the tree dies.

Pathogenic and parasitic fungi make use of various entry points into plants, including stomata (the openings on leaf surfaces that allow plants to breathe) and wounds. And, of course, with all this talk of enzymes decaying tough-to-digest lignin, it shouldn't surprise any gardener that some fungi can dissolve the cuticle and cell walls of the plant it is attacking. If you think this is difficult, think about the fungi that penetrate bathroom tile, and know that some fungi can penetrate granite in search of food.

This entire book could be filled with descriptions of fungi that get their nutrition at the expense of living plants. This is not our purpose—only that you realize that soil is loaded with fungi, a concept most gardeners readily grasp because of direct experience.

Functional overlap with bacteria

It should be obvious by now that in a healthy soil food web, fungi and bacteria shoulder much the same work and share many of the same functions. Like bacteria, some fungi produce vitamins and antibiotics that kill pathogens in the soil as well as in the human body. Remember penicillin, the most famous fungus-turned-antibiotic of all? In 1928, when English bacteriologist Alexander Fleming returned to his lab after his vacation, he found a fungus had contaminated a petri dish full of *Staphylococcus* bacteria. It ruined his experiment, but no bacteria were found growing near the fungus, and the world of medicine has never been the same.

Fungi, like bacteria, play crucial roles in the soil food web as decomposers, nutrient cyclers, soil structure builders, and beneficial symbionts, preventing as well as causing diseases. As well, their ability to impact soil pH makes them an important tool for gardening with the soil food web.

Chapter 5
Algae and Slime Molds

LGAE AND SLIME MOLDS are not related; we merely group them together because, while they have roles in soil food webs, they generally don't affect gardeners. That said, we hope we have already made the point that the soil food web is a community of organisms playing out a drama: when one or another character is removed, it may have significant consequences on how the play unfolds.

Algae

Algae are broadly defined as single-celled or thread-like photosynthetic organisms, including seaweeds and even giant kelp. Who hasn't seen algae in a pond, river, or lake, at the beach, or, if not there, on the glass of a fish tank? There are three kinds of algae: marine, freshwater, and terrestrial, the latter often living in soil, on or near the surface (where sunlight is available), not near roots. While most algae require very moist conditions, it is surprising to find some that grow in hot deserts and at the frozen poles—though even these still require a film of water to survive.

Although algae are closely related to bacteria on the tree of life, they are often thought of as primitive plants because they are photoautotrophic, meaning they take energy from the sun and produce their own food. Indeed algae, like plants, are primary producers, not dependent on the soil's organic matter or other members of the soil food web for their food needs as are bacteria and fungi. Moreover, algae lack the specialization that characterizes higher plants, and, unlike plants, they have no true roots, leaves, or stems and don't have a vascular (water- and food-conducting) system. The cell walls of all but the diatoms, a form of algae, do contain cellulose, and in this way they are like plants. The cell walls of diatoms are composed of silica covered with an organic skin that decays and disappears after the organism dies, leaving behind, in huge numbers, the silica skeletons that make up diatomaceous earth, a product familiar to many gardeners.

Most gardeners associate algae with bodies of water, not the raised bed or lawn, yet there you will find them if there is enough moisture—terrestrial algae require not only light but a film of water in order to survive. A teaspoon of soil may contain anywhere from 10,000 to 100,000 cells of green algae (phylum Chlorophyta), yellow-green algae (Xanthophyta), and diatoms (Bacillariophyta). At one time algae served as pioneer organisms, growing on moist rock surfaces and, when they died, combining with weathered rock, and air, and water to form early soils. In this important way, algae helped start the succession of life by providing necessary organic matter when there was no other.

Diatom skeletons, 445×. Image copyright Dennis Kunkel Microscopy, Inc.

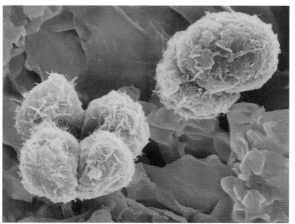

Green algae growing on the bark of a tree, 40×. Image copyright Dennis Kunkel Microscopy, Inc.

Algae help to create soil by forming carbonic acids as part of their metabolic functions. This causes rock to weather—a great example of chemical weathering brought about by biological activity. Resultant bits of minerals and the dead algae combine, producing soil eventually. This is not unlike the decay of rock surfaces caused by lichens—the symbiotic relationship between certain algae and fungi. The fungus provides a humid and somewhat protected environment in which the alga can live and, in return, receives photosynthesized food from the alga. In this relationship, the decay abilities of algae are aided by their fungal partners, and the process of weathering is sped up considerably. Lichens contribute nitrogen to the soil, and blue-green algae (Cyanophyta) use the enzyme nitrogenase to fix nitrogen, either in a symbiotic relationship or nonsymbiotically, similar to nitrogen-fixing bacteria. This is how rice plants can get nitrogen from the water in which they grow.

In truth, the role of algae in gardening is minor because of their need for sunlight, which can only penetrate a short distance into the soil. However, where they do exist in the soil, algae can excrete polysaccharides, mucilage, and slimes—all sticky stuff—which help bind and aggregate soil particles. Their presence can also help to form air passageways in otherwise compacted soil. And algae fit into some soil food webs as primary producers that are eaten by certain nematodes.

Slime molds

The slime molds are unusual-looking, amoeba-like organisms that inhabit damp, rotting wood, leaves, manure, lawn thatch, rotting mushrooms, and other organic material. They spend most of their lives pursuing bacteria and yeast in the soil. The few hundred different kinds of slime molds are in many ways like fungi but largely differentiated by the way they eat. Whereas fungi "digest" their food externally and then bring the nutrients inside the organism, slime molds engulf food and digest it internally.

The two groups of slime molds—Dictyosteliomycota (cellular slime molds) and Myxomycota (plasmodial slime molds)—have similar life cycles: they start out as spores and germinate into myxamoebae, amoeboid organisms that live in the soil and ingest bacteria, fungi spores, and small protozoa, locking up the nutrients they contain and preventing them from leaching out. They themselves are food for insect larvae, worms, and in particular, specialized beetles that have mandibles designed to scoop up the soft mold and cram it into their mouths.

At some point, for no apparent reason, individual myxamoebae swarm together; up to 125,000 or so form a mass that looks something like a big slug, a dollop of jelly, or, in some cases, vomit. These masses are of various sizes, in shades of tan, yellow, pink, or red, and are actually quite attractive in their own way. The species of one common plasmodial slime mold genus, *Physarum*, are usually about 1 inch (2.5 centimeter) thick and can grow to 1 foot (30 centimeters) or more wide.

Myxamoebae stage of slime mold on grass. Courtesy B. Clarke. Reprinted, with permission, from http://www.apsnet.org/, American Phytopathological Society, St. Paul, Minnesota.

Slime mold swarms can look like dog vomit. Photograph by Tom Volk, University of Wisconsin-La Crosse, www.TomVolkFungi.net.

The individual cells in the mass lose their walls, and the resulting plasmodium (or multinucleated mass of cytoplasm) emerges from the soil and slowly moves over leaves, grass, driveways, logs, mulch and anything else in its way. It does so at an average speed of 1 millimeter per hour, engulfing food as it goes. If a source of organic dead matter is put near a plasmodium, it will go to it. Even more amazingly, if you cut a plasmodium in half or even in quarters, the parts will come together again.

All sorts of theories have been put forward to explain why these organisms swarm. It may be that when food becomes scarce, there is a need to work together. There is, after all, something to be said for strength in numbers. What is known is that each individual myxamoeba leaves a bit of chemical attractant in its wake as it travels, presumably on a path toward nutrients. Other slime molds come in contact with this "slime sheet," not unlike that left by a slug as it travels, and take the same trail, adding their exudates to the path. As more and more organisms gather on the path, each adding its chemical slime to the mix, the attraction increases until literal swarms of myxamoebae congregate into a growing mass.

Eventually the plasmodium finds an appropriate spot and forms a fruiting structure, or sporangium. This unusual-looking body is of a distinctive shape for each slime mold species. Some sporangia are like tiny raised towers, in the top of which spores are formed. Sporangia come in yellow, blue, red, brown, and white and form a beautiful net of colors that really is as pretty as anything you can grow in your gardens.

From the soil food web perspective, slime molds help cycle nutrients, and the slime each individual myxamoeba creates helps bind soil particles. When conditions become unfavorable, plasmodia dry up and turn to powdery dust. Although these organisms don't play a major role in the garden, when a gardener comes across a slime mold, he or she remembers it.

Chapter 6

Protozoa

MOST GARDENERS first poked and prodded protozoa as part of a biology lab assignment, which invariably involved identifying and sketching cell parts of a paramecium; they may therefore remember that protozoa are single-celled organisms with a nucleus, which makes them eukaryotic and therefore, along with fungi, members of the domain Eukarya. Protozoa (which term we use descriptively in this book, as shorthand for a group of non-algal, non-fungal, animal-like unicellular organisms, across several kingdoms—but don't get us started!) are almost always heterotrophs, meaning they cannot make their own food. Instead, they obtain their nutrients by ingesting bacteria, primarily, but also the occasional fungus and, to a lesser extent, other protozoa.

Paramecia are still the favorite microbe. That's because these and other soil protozoa are considerably larger than bacteria, 5 to 500 micrometers versus 1 to 4 micrometers. This may still seem small to you, but in the scheme of microorganisms, 500 micrometers is pretty large—so large that under ideal lighting conditions a paramecium, at least in water, is visible to the human eye. You still have to look very carefully and certainly will not be able to differentiate any of those internal or external features you were told to label in school, but without a microscope you can see them flitting around. Through an electron microscope, unseen detail is observable.

Protozoa are something to stay away from if you are as small as a bacterium. By way of comparison, if a single bacterium was the size of a pea, a paramecium would be as large as a watermelon. This is why bacteria can hide from most protozoa in soil pores that are too small for the protozoa to reach into. Another way to make the comparison is to go back to that same teaspoon of good soil, with its billion-count bacteria—and "only" several thousand protozoa.

Over 60,000 kinds of protozoa are known and, contrary to any residual youthful hope you may have that they only live in pond water, a majority of them live in the soil; however, all do require moisture to lead an active life.

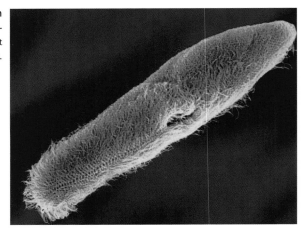

Given the crucial role protozoa play, a quick review of some school biology—and then some—is in order.

Amoebae, ciliates, and flagellates

Protozoa come in three basic "models." First are the pseudopods, single-celled animals with amorphous forms most will remember as amoebae. These are constantly on the move, a feat (if you will pardon the poor pun) accomplished by pouring their cytoplasm—the soup with all its life parts—into one or more false appendages called pseudopodia ("false feet"). Pseudopods themselves are of two types. The first has a shell-like exoskeleton and five predefined holes (think of a bowling or golf glove), through which the pseudopodia can appear. The other class lacks any shell or predefined pseudopodia; these amoebae are relatively large microorganisms, and many would be as visible as paramecia if they weren't so transparent. Amoebae lack a mouth and ingest bacteria by surrounding them and engulfing them in gas bubbles, into which are transmitted digestive enzymes. The entire vesicle is then absorbed, and waste products subsequently expelled.

Next in size are the ciliates. These protozoa are considerably smaller than their amoeboid cousins but still much larger than their bacterial prey. Ciliates are covered with rows of hairs that beat like the slaves' oars on a Roman galley, propelling the organism to food—or away from enemies. In addition, these "oars" create currents that bring bacteria into the ciliate's mouth region, so they can be ingested. The familiar paramecium is a ciliate protozoan.

The third and smallest type of protozoa are the flagellates. Their one or two long, whip-like hairs, or flagella, allow them to move about in search of food.

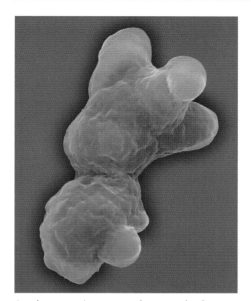

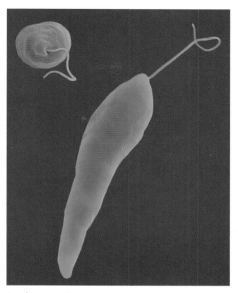

An electron microscope photograph of an amoeba, 700×. Image copyright Dennis Kunkel Microscopy, Inc.

Euglena, 440×. Image copyright Dennis Kunkel Microscopy, Inc.

A few flagellates, like euglena (the "classic" freshwater flagellates of pond water), produce their own food via photosynthesis and are thus autotrophs; most, however, are heterotrophic, obtaining nutrients from eating and digesting other organisms in the soil.

More symbiotic relationships

As so many of the soil food web organisms do, protozoa form symbiotic relationships, particularly with bacteria, to such an extent that such associations appear to be the norm rather than the exception. A classic example is that of the flagellates residing in the guts of termites, which digest the wood fibers the termite eats. We now know that the relationship is actually a three-way one. Electron microscopy reveals working bacteria in the gut of the termite as well; these fix nitrogen from the atmosphere for the flagellates. Not often do you find a triple symbiotic relationship, though surely more will be discovered as exploration courtesy of the electron microscope continues.

Lots of ciliates, too, enter into symbiotic relationships with bacteria. Some ciliates live in sand and "farm" bacteria colonies; and it is the methane-generating bacteria inside ciliates that are responsible, in part, for the methane gas that develops in some ciliates as anaerobic respiration takes place.

Population-control police

Protozoa are attracted to and enter into an area where there is a good supply of bacteria in a pretty consistent progression (on average, a protozoan can eat 10,000 bacteria a day). First come the flagellates, the smallest of these microbes; these can move into small spaces in the soil, places where the large protozoa cannot and where bacteria are plentiful. Even after the larger ciliates arrive on the scene, the still-large population of bacteria provides plenty of sustenance for both the original flagellates and the newer ciliates. Finally, amoebae move through in search of bacterial prey (and also smaller protozoa). The combined pressure on the bacterial population becomes so great, numbers start to diminish. As readily available bacteria become harder to find, the larger ciliates and amoebae start to eat more of the smaller ciliates and flagellates. This reduces the population of ciliates and flagellates which, in turn, allows the populations of bacteria to stabilize and return to a level that maintains the soil food web balance.

Why aren't all the bacteria consumed by protozoa? One reason is that protozoa are restricted by bacterial slime; this film is hard for them to penetrate, and it lacks the oxygen that they require. Another reason is that the bacteria are smaller and able to hide in tiny soil pores.

It seems counterintuitive that increasing protozoa populations most often results in increases in the bacterial populations upon which they prey. This occurs because fewer bacteria means less competition for nutrients among the surviving bacteria. Not having to compete all the time for food means they can divide well fed. Likewise their progeny will have something to eat so they can multiply as well. If protozoa can keep their own numbers in check, they have all the bacteria and fungi they need to eat.

It is not just populations of bacteria that protozoa keep in balance. In their search for sustenance, some protozoa attack nematodes. Others reduce nematode populations by competing for the same, limited food resources, i.e., other protozoa and fungi. This also helps keep bad-guy nematode populations from flourishing.

Protozoa need moisture to live, travel, and reproduce, and hydroscopic water—that thin film of water left on the surfaces of soil particles and aggregates—provides it, under normal soil conditions. If things dry up, however, most protozoa stop feeding and dividing and go dormant, encasing themselves in a cyst. How long protozoa can survive in this state varies from species to species; some can withstand an extended dry spell of several years. This technique ensures the survival of both the protozoa and the plants that benefit from the nitrogen and other nutrients released by their activity.

Mineralizers

Of critical importance to the workings of the soil food web are the waste products produced when protozoa ingest bacteria or fungi. These wastes contain carbon and other nutritional compounds that had been immobilized but are now once again mineralized and made available to plants. Nitrogen compounds, including ammonium (NH_4^+), are among them. If nitrogen-fixing bacteria are present (remember, these usually require a pH of 7 or above to have good populations), free ammonium is converted into nitrates. If not, the nitrogen remains in ammonium form.

Mineralization of nutrients is crucial to the survival of plants in a natural system. Our premise is that by interfering with or destroying the soil food web, the gardener has to step in and do extra work, making gardening a chore instead of an enjoyable hobby. If you are not convinced, then consider that as much as 80% of the nitrogen a plant needs comes from the wastes produced by bacteria- and fungi-eating protozoa. Since bacteria and fungi are attracted by plant exudates to the rhizosphere, and that is where protozoa consume them, a huge source of plant food is delivered, right around the roots.

Other soil food web functions

All protozoa participate to some degree in the decay process by inadvertently ingesting small particles of organic matter. These are then broken up into smaller pieces if not totally digested and become available to bacteria and fungi in the waste stream. And other soil food web members rely on protozoa as one of their food sources—another reminder that it is a soil food web, not chain, with which we are dealing. Certain nematodes, for example, are dependent on protozoa as their food source and have developed specialized mouthparts to better ingest them. Worms too rely on healthy populations of protozoa. Without protozoa in the area, gardens are devoid of worms. Similarly, many microarthropods require a healthy dose of protozoa to thrive.

Finally, not all protozoa are beneficial. Several kinds eat roots, but in a healthy food web these are kept in check by other, cannibalistic protozoa. So to a certain degree protozoa serve as a food source for themselves—and remain, even the worst of them, crucial characters in a healthy soil food web.

Chapter 7

Nematodes

NEMATODES are nonsegmented, blind roundworms that, along with protozoa, mineralize nutrients contained in bacteria and fungi. Their name is derived from the Greek *nema*, which means "thread," an apt descriptor for these microorganisms. Nematodes are considerably larger than protozoa, with lengths averaging 2 millimeters and diameters of 50 micrometers (versus 0.5 millimeter for a decent-sized protozoan). Still, most nematodes are difficult to see without a microscope. When you can see them with the naked eye, they usually look like moving human hairs. We say "usually" because the largest known nematode, *Placentonema gigantissima*, can grow to a showstopping 30 feet (9 meters). Fortunately, this nematode lives in the placenta of sperm whales, not in soil.

These fascinating roundworms are actually the second most dominant form of animal life next to the arthropods. Over 20,000 species of nematodes have been identified thus far, and scientists suggest there may be as many as 1 million species in total. They are everywhere, yet most gardeners know little about them save the parasitic ones that damage roots.

Our teaspoon of good soil teeming with microbial life averages about 20 bacteria-eating nematodes, 20 fungal feeders, and a few predatory and plant-eating nematodes, making the total number between 40 to 50 nematodes. The number of fungi- versus bacteria-eating nematodes is directly related to the availability of the food sources they require.

Picky eaters

Nematodes are major consumers in the soil. All have a long alimentary tract that runs from the mouth to the anus, which is located in their tail end. Nematode skin is actually a cuticle; it protects the animal from both physical and chemical attack and provides lightweight structural support. For the gardener, the best way to classify them is by their eating habits: various nematodes have developed specialized mouthparts to allow them to attack and get at their own particular brand of prey.

Let's start with the nematodes that eat living plant material. These plant parasites usually have needle-like stylets that enable them to puncture plant cell walls with ease. Some of these root-eating nematodes are ectoparasitic (that is, they feed on the root surface), while others are endoparasitic, entering into the root to feed. Herbivore (plant-eating) nematodes can create lesions in the root as well as cysts and large bulges that gardeners refer to as root knots. Obviously, nematodes that eat roots do not help the crop in question.

Next are bacterivores, nematodes that eat bacteria. Here the specialized mouthpart is usually a hollow tube. A bacteria-eating nematode so equipped can consume multitudes of tiny bacteria in an hour. Other nematodes are fungivores: they eat fungi. This type of nematode also has stylets, for puncturing the chitin cell walls of fungal hyphae. Like their fellow fertilizer-spreaders, protozoa, both these types of nematodes mineralize the nutrients contained in the smaller microbes' bodies, making them once again plant-available.

Predator nematodes feed on protozoa, algae (including diatoms), and other small members of the soil food web—grubs, weevils, wasps, even small invertebrates such as slugs (the first beneficial nematodes sold for gardening use was to control slugs). Predatory nematodes eat other nematodes as well, thus preventing overgrazing of bacteria and fungi and keeping populations of

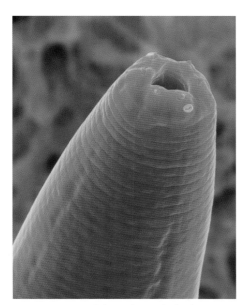

An SEM image of the stylet end of a fungi-eating nematode. Image copyright Dennis Kunkel Microscopy, Inc.

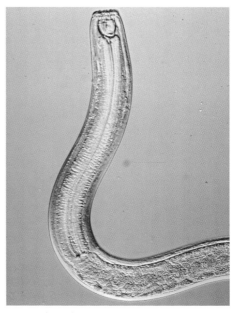

A typical predatory nematode. Photograph by Bruce Jaffee, UC Davis.

destructive nematodes, primarily the herbivores, down. As with protozoa grazing on bacteria, fungal grazing by nematodes frees up fungal food resources to such an extent that fungal populations are increased, to a point, resulting in increased decay of organic litter. Thus nematodes are indirectly responsible for some of the decay that occurs in and on the soil.

Other nematodes are omnivores, eating any and all of the above, to the merest fungus spore. Some even ingest organic matter and so are directly responsible for the decomposition of organic matter.

Mineralization and other tricks

Arguably, mineralization is the most important thing nematodes (at least the bacterivores and fungivores) do for gardeners. Nematodes need less nitrogen than protozoa do; those that eat fungi and bacteria, therefore, release even more of the previously immobilized nitrogen into the rhizosphere, in ammonium form. Again, if the populations of nitrogen-fixing bacteria in the area are low (as they will be when the pH is below 7), the mineralized nitrogen remains predominately in ammonium form (that is, it is not converted to nitrate).

But here's something new. Because nematodes are bigger than bacteria, fungi, and protozoa, they require more porous soils in which to travel, and their numbers will be reduced if the soil is of the wrong texture or if it is too compacted. Either of these conditions will block nematodes from searching for nutrients. Unable to search for food, they either die off or move elsewhere, and the flow of nitrogen available to plants is greatly diminished.

Not just these escapees but all nematodes inadvertently play a role in transporting bacteria to areas far from their origin. This is because bacteria attach to the skin of nematodes and are spread to other areas as the nematode makes its way through the soil in search of food. Since bacteria themselves have extremely low mobility in the soil, this is a great advantage to them: they can "taxi" to new food sources. It also may be said to help the nematode, who in the long run occasionally eats the progeny of its fare and increases mineralization in a new area. Fungi, too, can hitch a ride on a nematode. Often this is because the hapless nematode is prey to a fungus attack and goes about its business while it is being eaten alive.

Nematodes have developed some interesting ways to locate food in the soil. They may have specialized mouthparts, but they do not have eyes. How does a blind nematode survive in the soil or anywhere else for that matter? Some nematodes can sense extremely minute variations in soil temperatures. They "know" what temperatures particular food sources live in; they will move

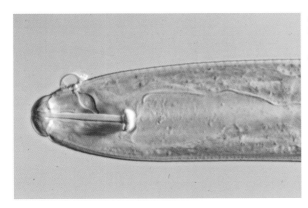

A fungal spore and tube has entered the side of this nematode and is heading toward its retracted stylet. Photograph by Bruce Jaffee, UC Davis.

through soil until they find the right temperature gradient and continue to travel along it until they bump into their preferred food.

Others find food by sensing particular chemicals associated with them. Once they are on the scent, they act like heat-seeking missiles, locking in on their prey and attacking. Our favorite fungus, the one that captures nematodes in its rings, attracts nematodes with a chemical. Clearly, there are disadvantages to this method of finding food.

In the end, nematodes are incredibly diverse and interesting animals that, like each and every other organism in the soil food web, deserve (and have) their own books.

Chapter 8

Arthropods

EVEN IF YOU DIDN'T KNOW what they were called, you've seen and know lots of arthropods: flies, beetles, and spiders, for example. No exaggeration: arthropods rule the world. Somewhere around three-fourths of all living organisms are arthropods. Still, for all their numbers and their larger size, the arthropods do not lead in terms of biomass: the biomass of nematodes and even protozoa is greater.

Arthropod is Greek for "segmented feet" (actually, arthropods possess segmented limbs and segmented bodies, but you get the idea). In addition to jointed legs and segmented bodies, all arthropods have in common an exoskeleton made from chitin, the same material that makes up the walls of fungus cells. You are familiar with the shells of lobsters, shrimps, and crabs, familiar examples of marine arthropods; their shells are chitin. As with the cuticle-skin of nematodes, this exoskeleton provides protection and a lightweight structural frame (an internal skeleton is considerably more complicated and heavier). As arthropods grow, they shed their exoskeleton and grow a new, larger one.

Arthropods usually have three (but may have only two) body segments, starting with a head, or cephalum; then a chest, or thorax; and finally an abdomen. Most arthropods live life in three stages. They start as eggs; hatch and live the early part of their lives in a larval form; and then metamorphose into a very different form for their adult lives. A caterpillar, to use one famous example, is the larval stage of a butterfly, the adult who will lay eggs to start the cycle over again. Many arthropods live all three stages in or on the soil, but many reside there during just one or two. Of course, any gardener who has fended off cutworms fully appreciates that it takes only one stage to damage a plant.

Arthropods range in size from the humongous Alaskan king crabs that measure a couple of yards across to tiny mites that require a powerful microscope to see. Those that can be seen only under magnification are classified as microarthropods; those that can be seen easily without the aid of a hand lens or microscope are known as macroarthropods.

Besides being food for other members of the soil food web, soil arthropods are important to the community as shredders, predators, and soil aerators. The presence or absence of certain of these key players can tell a gardener much about the health of soils and the plants growing in them.

Classifying arthropods

Without more than a passing interest in them, most gardeners lump all the arthropods together as simply "insects" or "bugs." Any given gardener may know a few of the popular and unpopular ones that inhabit local gardens, but for the most part, not many more. Part of the problem is that there are too many arthropods: the phylum Arthropoda is by far the largest in the animal kingdom—so large that it presents a real challenge for us: how can we the authors show you the readers how to use the soil food web without overwhelming you with information? There are just too many kinds of soil-dwelling arthropods to describe them all, or even come close to doing so, and frankly, there is too much scientific nomenclature as well. Bear with us for the little we do use.

Gardeners are agreed that using scientific names, usually derived from Latin or Greek, is truly the only way to accurately identify a plant; but most have not learned the alphabet soup of words scientists use to classify members of the phylum Arthropoda, whose members have the greatest impact on the soil food web. Here, we list the classes, as a start:

Class Arachnida: spiders, scorpions, mites, ticks, and daddy longlegs

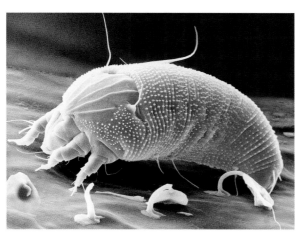

A rust mite (*Aceria anthocoptes*), 700×. Photograph by Eric Erbe, digital color by Christopher Pooley, USDA-ARS.

Dust mites (*Tyrophagus putrescentiae*), 100×. Photograph by Eric Erbe, digital color by Christopher Pooley, USDA-ARS.

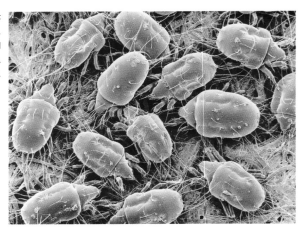

Millipede foraging on soil. Photograph by Frank Peairs, Gillette Entomology Club.

Female Mormon cricket. Photograph by Michael Thompson, USDA-ARS.

Class Chilopoda: centipedes

Class Diplopoda: millipedes

Class Insecta: springtails, silverfish, termites, mayflies, dragonflies, damselflies, stoneflies, earwigs, mantids, cockroaches, walking sticks, grasshoppers, katydids, crickets, rock crawlers, web spinners, zorapterans, psocids, book lice, bark lice, chewing lice, sucking lice, scorpion flies, fleas, thrips, lacewings, ant lions, true bugs, moths, butterflies, flies, beetles, sawflies, bees, wasps, and ants

Class Malacostraca: sow bugs and pill bugs

You are already familiar with many members of the class Insecta. Tens of thousands of different kinds of insects live in and on the soil and plants, as few gardeners need to be reminded. Surely you have seen representatives of one order of this one class, the order Coleoptera (beetles), as you go about your garden-

The predatory beetle *Thanasimus formicarius* feeds on the pine shoot beetle, a serious pest of pines. Photograph by Scott Bauer, USDA-ARS.

Formosan subterranean termites feeding on spruce and birch wood. Photograph by Peggy Greb, USDA-ARS.

ing chores: with approximately 290,000 species described, it would be hard to miss them.

Soil food web functions

Most soil arthropods, particularly those that reside on the soil surface, are shredders. They chew up organic matter in their constant quest for food, creating smaller pieces. As a result, fungal and bacterial activity is increased because shredding exposes surfaces on organic litter that give bacteria and fungi an easier avenue of attack.

As they shred and move about, arthropods also taxi microbial life attached to their bodies or in the debris they push or carry about. Since most arthropods are food for still larger animals, the total distances microbes can be moved (consider a bacteria colony eaten by a grub that is then ingested by a robin) can be truly great. Microbial activity is increased if the taxi takes its fare to a good food source. Still, it is the shredding that is most important. Two common arthropods, mites and springtails, are alone responsible for recycling up to 30% of the leaves and woody debris deposited on a temperate zone forest floor.

In the face of insufficient dead organic matter, arthropods often attack living sources of organic nutrients. And even if the supply of available organic matter is abundant enough to satisfy any reasonable arthropod, some (mole crickets, root maggots, cicadas) subsist on roots anyhow. Fungus gnat larvae, for example, hatch and immediately start eating root hairs, eventually eating their way into the roots and stem to the great detriment of the invaded plant.

Dark-winged fungus gnat larvae.
Photograph by Whitney Cranshaw,
Gillette Entomology Club.

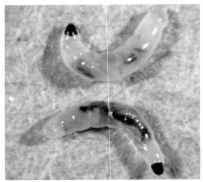

Still other arthropods eat other members of the soil food web in order to survive; by removing their fellows, these predator arthropods make room for other arthropods to fill the emptied niche, helping to create complete digestion of soil matter. Finally, in much the same way protozoa and nematodes do, some arthropods eat fungi, others bacteria, but this time releasing nutrients on a larger scale, befitting their greater numbers and size.

Many arthropods carry out their daily routines only on the surface of the soil. A surprising number, however, live at least part-time below the soil surface. As these arthropods go about their business, they mix and aerate soil; their waste products also add organic matter.

Mites

Several soil arthropods play dominant roles in the soil food web. Among them are mites, of which there are two basic kinds in the soil. The first, oribatid mites, actually have the highest populations of any soil arthropod, with up to several hundred thousand per square yard; a primary reason for this is that the female oribatid mite doesn't need a mate to lay fertilized eggs. These important mites are 0.2 to 1 millimeter long. Oribatid mites inhabit soil surfaces, particularly litter debris but also live plants, including mosses, and lichens. Some oribatid mites feed on live nematodes, others on dead springtails. Most, however, eat fungi and algae and decaying plant matter and, because of their large numbers, are major recyclers and decomposers in the soil food web. Although they are vulnerable when they are born and in the later nymph stage, as adults their exoskeletons make these mites impervious to most forms of predation except by ants, beetles, and larger animals, like salamanders.

The second kind of soil mites, gamasid mites, are major predators in the soil. Their populations (and there can be several hundred gamasid mites in a square yard of soil) are dependent on the availability of their food source, which happens to be most any other arthropod that grazes in the soil. As such, the presence and numbers of gamasid mites are considered useful tools in determining soil health: if there are lots of them, their fellows must be plentiful, and that usually means a healthy soil food web. Soft-bodied for an arthropod, they fare less well than oribatid mites against predators, however, and are themselves prey to all sorts of other arthropods.

Most gamasid mites act in a manner reminiscent of spiders (with whom they are often confused—all mites, like spiders, have eight legs): they inject their victims with enzymes that dissolve their innards and turn them into a liquid the mites then suck. Gamasids subsist on collembolans, insect larvae, and insect eggs. Those that live in the soil, as opposed to on its surface, also eat nematodes and fungi.

Springtails

The springtails (*Collembola* spp.), another important group of arthropods, are among the more active insects in the soil. You can expect to find up to 100 of these "soil fleas" per square inch in soils with enough organic material. Ranging from 0.2 to 2 millimeters long, they are often seen as little critters that jump into the air when the top layer of soil debris is disturbed.

Springtails lack wings. Instead, they possess a forked tail, or furcula, that folds beneath them but is capable of being straightened in an instant (fluid pours into its base), propelling the animal up to a yard backward (hence their common name), out of harm's way.

As with many members of the soil community, springtails have adapted to several different kinds of environments. Those dwelling on the surface, for

Springtail showing the well-developed furcula that enables these animals to "jump" up to a meter away from predators. Photograph by Michael W. Davidson, Florida State University.

example, have well-developed furculas, eyes, long legs, and antennae, while those that dwell in deeper parts of the soil are blind or near blind and don't need a large furcula or long legs, as these would be hindrances in traveling about in search of food. Some springtails, with even more developed "springing tails," are especially adapted to live among grasses.

A springtail's diet consists of bacteria, fungi, and decaying organic matter. Springtails are also sometimes consumers of nematodes and dead animal matter, and are themselves the favorite food of mites.

Termites and ants

Two other abundantly represented members of the soil food web, termites and ants, are not really related even though they look similar. Ants are related to bees and wasps; they usually have eyes, opaque bodies, a narrow waist, long legs, and a hard exoskeleton. Termites, by contrast, are blind and have soft, translucent bodies and short legs. The shredding activity of both these insects helps decompose organic matter on the soil surface.

Termites eat mostly materials containing cellulose. As do other arthropods, they open up organic matter, making it easier for fungi and bacteria to get at it. Some of this organic matter is brought down into tunnels and burrows, where it becomes available to different populations of microbes. Indeed, it is the construction of tunnels and mounds that distinguish both termites and ants from other microarthropods. In constructing their homes, ants and termites mix surface and subsurface soil. In the case of ants, up to six tons of soil may be mixed a year. In tropical areas, the contribution ant and termite activity makes to the mixing of soil is greater than that of worms. Ant and termite tunnels obviously provide a way for air and water to get into the soil and for other animals to move about. Sometimes these tunnels make it easier for roots to penetrate the soil; often, roots will follow tunnels.

Termite and ant mounds formed on the surface of soil contain subsurface material, and as these mounds are weathered and break apart, they alter the surface soil mix. Finally, termite hindguts contain anaerobic bacteria that produce methane, so much so that termites are a major contributor of this greenhouse gas to the atmosphere.

In sum, because of their huge numbers and the varied jobs they perform, micro- and macroarthropods are crucial to any functioning soil food web, and their presence, both in numbers and kind, is an indication that the community is not only working but healthy and thriving.

Chapter 9
Earthworms

ARTHWORMS are the most recognizable of all animals in the soil food web and, as it turns out, one of the most important to gardening. Most probably the ones you will run into will be a species of *Aporrectodea*, *Eisenia*, or *Lumbricus*, unfamiliar generic names for the most familiar of the 7000 or so species of earthworms common to good garden soils. Technically earthworms are segmented worms, or oligochaetes, and grow anywhere from a few inches to as much as a yard in length. They include the smaller, less familiar pot worm (*Enchytraeus doerjesi*) found in forest soils (gardeners may not be familiar with pot worms unless they have tropical fish, for which live pot worms are a favorite food). Pot worms are much smaller than the traditional garden earthworm, only a few millimeters to a few centimeters in length; they succeed and replace earthworms in acid forest soils, which earthworms shun. As unbelievable as it sounds, an acre of good garden soil contains 2 to 3 million earthworms (anywhere from 10 to 50 per square foot); this is enough to do a bulldozer's amount of work and indeed, this crew is capable of moving an astonishing 18 tons of soil a year in search of food. In an acre of forest soil, one might find about 50,000 of their cousins—a large number, but small in comparison. Obviously, earthworms do not play as large a role in the soil food web of forests as they do in that of gardens.

Early European settlers transported many earthworm varieties to the eastern coast of North America. The worms rode along in potted plants and ship ballast and, one would imagine, arrived as valuable luggage cherished by farmers, who knew the high worth worms would have in the new world. Once here, they moved across the continent in soils that held fruit trees and other nursery stock. They thrived. The only place in North America European worms have not done well is in the warm desert of the Southwest. The common night crawler (*Lumbricus terrestris*), for example, which is dominant in garden soils from sea to shining sea, arrived with the Europeans. Nor is the red wiggler (*Eisenia fetida*), a common compost worm, a native (though it is often called the Wisconsin red wiggler); still, it is a favorite (and rightly so) with those that maintain vermicompost (worm composting) bins. All earthworms have the ability to spread into new areas, survive, and multiply to tremendous populations.

It takes two worms to produce offspring, although worms carry both sets of sexual organs. Each has a slime tube in which to incubate eggs that are placed in a small cocoon. Each cocoon contains 15 or more baby worms, who themselves, once hatched, are usually mature enough to breed in only three or four months. When one considers that some worms live for 15 years, breeding all the while, their large populations in soil are understandable.

Worms are a powerful force in the soil. Charles Darwin, who studied them at length (and even wrote a book about them, *The Formation of Vegetable Mould Through the Action of Worms with Observations on Their Habits*), argued that every particle of soil has been through a worm at least once. Whether he was right or wrong, their role in the soil food web is key. They are intimately involved in the shredding of organic matter, the aeration of soil, the aggregation of soil particles, and the movement of organic matter and microorganisms throughout the soil. They also increase microbial populations and aid plant root growth.

Eating machines

Although earthworms have no eyes, sensory cells in their skin are very sensitive to light. Their mouth, or prostomium, is a fleshy pad that looks somewhat like an extended lip; it, along with the worm's pharynx, is extremely muscular, but there are no teeth.

What does a worm eat? Bacteria, primarily, which is why it should come as no surprise that soils with large populations of worms are usually bacterially dominated. Other foods are fungi, nematodes, and protozoa, as well as the organic matter on or in which these microorganisms live. How does a worm eat?

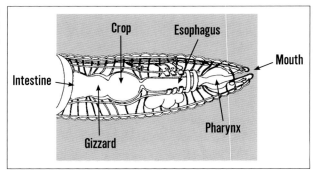

The business end of an earthworm. Courtesy Tom Hoffman Graphic Design.

It starts by pushing its pharynx out of its mouth and uses it and its prostomium to grab food and bring it inside its body. In the food goes, and strong muscles start to break it down into particles. Saliva is mixed in, moistening things up.

Next, the food travels down the worm's esophagus into a crop. From this storage compartment it travels to the gizzard, an extremely strong muscle that is partially filled with sand and small rock particles. As the gizzard contracts and extends, the food is ground up by the sand, which serves as the "teeth" of the toothless worm. When the food is sufficiently ground up, it travels to the worm's intestine. Just before, however, it is mixed with a liquid calcium carbonate.

Given their reputation for recycling organic matter, it is surprising to say the least that earthworms lack the necessary enzymes to digest it, relying instead on bacteria. All the grinding in the gizzard assures that what food does arrive in their intestines is sufficiently small and opened up for the bacteria living there to quickly digest it. Nutrients produced by the bacteria are finally absorbed into the worm's bloodstream, and any organic matter that is not fully digested is eliminated. This may be useless waste to a worm, but to a gardener it is a fantastic soil amendment.

Vermicastings

Vermicastings (the name given to worm poop) are 50% higher in organic matter than soil that has not moved through worms. This is an astonishing increase and radically changes the composition of the soil, increasing CEC because of the greater amount of charge-holding organic surfaces. Other nutrients, therefore, have the ability to attach to the organic matter that has passed through a worm.

The benefits don't stop there. The worm's digestive enzymes (or, properly, those produced by bacteria in the worm's intestines) unlock many of the chemical bonds that otherwise tie up nutrients and prevent their being plant-available. Thus, vermicastings are as much as seven times richer in phosphate than soil that has not been through an earthworm. They have ten times the available potash; five times the nitrogen; three times the usable magnesium; and they are one and a half times higher in calcium (thanks to the calcium carbonate added during digestion). All these nutrients bind onto organic matter in the fecal pellets.

Worms can deposit a staggering 10 to 15 tons of castings per acre on the surface annually. This almost unbelievable number is clearly significant to gardeners: the ability to increase the availability of nutrients without carting in and adding tons of fertilizer is about as close to alchemy as one can get.

Master shredders

Earthworms are classified as shredders. As they search for food, they break down the leaf litter in the garden and on the lawn, greatly speeding up the decomposition of plant material, directly and indirectly. They open up leaves and other organic matter, giving bacteria and fungi better access to the cellulose (and other carbohydrates) and lignin (a noncarbohydrate) in the organic matter. Earthworms, then, obviously facilitate the recycling of nutrients back to the plants. At the same time, they may also change the composition of the food web community by competing with fungi and bacteria for nutrients, indeed even eating into their very populations. The magnitude of the impact of earthworms is shown by this simple fact: leaves on the forest floor or in a garden or lawn would normally require one and perhaps two years to decay without worm shredding, but only three months with it. In some parts of the United States and Canada, forests have been invaded by earthworms left by fishermen. These have completely altered the floor habitat, and entire forests are affected as the litter layer is being decayed far faster than is healthy for the trees and the rest of the soil food web.

The end results of worm shredding and digestion are minute particles of organic litter that microorganisms can eat. Microbial populations in the soil are also enhanced because some microbes are mixed into worm fecal pellets during their formation and elimination, creating protected enclaves of fungi and bacteria.

A worm leaving its castings on the surface of a lawn. Courtesy USDA-NRCS.

Burrows

Earthworms are incredibly strong, a necessity given the amount of burrowing they do. While making their way through the soil to feed, worms can move rocks that are six times their weight. Being in the soil provides them with moisture, temperature control, and protection from birds and other aboveground predators.

Different kinds of worms make different kinds of burrows, some permanent and others temporary. The temporary burrows are often abandoned after they become filled with castings and litter; roots grow into these pathways, able to penetrate deeper than they could by themselves, all the while having access to nutrients and the microorganisms that freed them. Certain kinds of earthworms move up and down in the soil, sometimes as deep as 12 feet. They shred litter on the surface and pull some of it into their burrows, where it is later decomposed. In making tunnels, soil from deeper in the ground is deposited on the surface. Other earthworms travel horizontally, rarely leaving the top 6 inches (15 centimeters) of soil, but even these redistribute organic matter several feet (a meter or so) away, though in the same horizon. Either way, this movement is akin to delivering food to another area of town and impacts the entire population of a soil food web. Earthworms also move microorganisms, whether attached to their own bodies or to the litter they pull underground, starting communities where once there were none.

Earthworms not only increase a soil's porosity, but by breaking down and mixing organics, they also increase its water-holding capacity. Again, think of a couple of million worms burrowing about in that acre of good garden soil. Their burrows become significant pathways for water drainage and air passage. And since some worms move vertically and some horizontally through the soil, these pathways can bring water to all sorts of underground locations, whether put to immediate use by plants or stored, for later absorption.

Everyone loves earthworms

Other than birds, a few parasites and parasitic flies, and the occasional mammal (a mole, a fisherman—a tropical-fish fancier), earthworms have few enemies. The birds they attract to the lawn eat them, but from a soil food web perspective, all is not lost. Not only does bird guano contain nutrients and microorganisms, but bird feet carry protozoa, and these are spread about when the bird hops from spot to spot. And, occasionally, a bird will drop a worm into a new location (but not the early bird, who always *gets* the worm).

Look at the benefits of earthworms. They shred debris so other organisms can more readily digest them. They increase the porosity, water-holding capacity, fertility, and organic matter of soils. They break up hard soils, create root paths, and help bind soil particles together; they cycle nutrients and microbes to new locations as they work their way through soils in search of food. With all these benefits, isn't it strange to count the gardener as one of the predators of the earthworm? Rototilling and other mechanical methods of turning soil destroy worm burrows and reduce or even destroy earthworm populations by cutting them up into pieces that don't ever regenerate whole worms. And the gardener who uses chemical fertilizers is literally throwing salt on the wound: these chemicals are salts that irritate worms and chase them out of garden soils.

A noticeable worm population is a clear sign of a healthy food web community. It means organic matter, bacteria, fungi, protozoa, and nematodes—all necessary to support a worm population—are in place. With these at the base, chances are the other parts of the soil food web are in order as well.

Chapter 10

Gastropods

W
HAT GARDENER hasn't had a run-in with certain members of the order Mollusca? Or perhaps you know them as slugs and snails. These gastropods (Greek for "stomach-foot") are often called mollusks, but whoever bestowed the common name with Greek roots had the right idea—it is an apt description of what these organisms seem to be: one big foot that does a lot of eating. Most garden slugs are the size of a fingernail, but some species grow to 18 inches, fulfilling every gardener's nightmare. Besides, mollusks are usually associated with salt- and freshwater creatures, clams and oysters in particular, not the garden. With some 40,000 species, gastropods are the largest group in the order Mollusca.

Land snails, from which slugs further evolved, emerged from the sea some 350 million years ago complete with the shells developed to protect them from their water-dwelling enemies and chemicals in salt water. As one would expect from their appearance and the damage either can inflict to a garden, both slugs and snails have a similar physiology. The main difference between the two is the snail's shell, which is made of calcium. Garden slugs evolved from these snails over the years and, depending on the species, either entirely or mostly lost the shell.

Slugs and snails are extremely susceptible to dehydration. Here is where the snail has an advantage over the slug. The slug must find cover in a moist area to survive dry times. A snail can pull into its shell; seal off the opening by secreting slime material that hardens into a thick, leathery layer, or operculum; and remain up to four years inside its sealed-off shell. When the snail is ready to emerge, it simply eats its way through the operculum and is good to go.

Why would snails evolve to slugs, losing such a wonderful device as a shell? Not having a shell has its own distinct advantages. Clearly a slug has greater mobility and shape control; it can squeeze into spaces a hard shell would not permit, vastly increasing its scavenging range (reported to be up to a mile a night). In addition, maintenance of a shell requires access to calcium, limiting the areas in which any particular shelled gastropod can live. Slugs, who need less calcium, are under no such stricture; their freedom to roam and ability to get to new sources of food is unimpeded.

Garden slugs and snails are nocturnal, probably because this is the time of highest moisture or the least amount of drying heat. It may also be that this is when they are less susceptible to predators. They spend the day hiding in the soil or under debris. When nightfall arrives, they move about by gliding on a single, muscular "foot" through which they secrete glycoproteins—a sticky slime of sugars and proteins.

This slime is manufactured in cells located in the snail's or slug's muscular foot and is exuded from the foot's center. The outer edges of the foot then stretch and slide forward over the slime. Slugs and snails are able to stretch up to 20 times the length of their bodies. The lubricant later hardens to form paths recognizable by other snails or slugs (and gardeners) or by the same slug returning from foraging. Amazingly, the slime contains chemicals obnoxious to predators, just in case the gastropod is being followed.

Slugs and snails are hermaphrodites, meaning they are capable of self-fertilization; most cross-fertilize, however, allowing both mates to lay 100 to 200 translucent oval eggs up to six times a year. These are deposited just under the soil surface, where they can remain for years until conditions, primarily moisture, are just right. They hatch in as little as two weeks, however, if conditions are right—which they usually are in the garden. Slug and snail juveniles are tiny, but they are ready to eat like adults and search for food a day or two after birth. They return to their "nest" each morning for the first several

A red slug, foraging. Photograph by Gary Bernon, USDA-APHIS, www.forestryimages.org.

months; they become sexually mature after about six months and are fully grown after about two years.

You might think they are eating only your lettuce and kale crops, but both snails and slugs also graze on fungi, algae, lichens, and rotting organic matter. Believe it or not, they don't graze only on surface plants. It has been reported that slugs spend a mere 5 to 10% of their time above ground. For every slug you see above ground, three or four are underground, foraging in the soil. Both snails and slugs possess a radula, a series of chitinous teeth not unlike a wood rasp, which allows these garden gastropods to grind their food down to very tiny particles. Many slugs and snails are capable of digesting cellulose.

Snails and slugs have a place in the soil food web. They speed decomposition and decay by shredding their food before they consume it. Like earthworms and some of the arthropods, they open up organic matter so that fungi and bacteria can get at it. Their underground travels create pathways for air, water, and roots; the slime they produce helps bind soil. They themselves are a food source for ground and rove beetles (particularly in their larval stages), spiders, garden snakes, salamanders, lizards, and birds. Some nematodes that subsist on slugs are now available commercially; these blind worms "heat-sink" in on a hapless slug, parts of which become a meal for the successful nematode while the remainder is left to bacterial and fungal colonization and decay.

When gastropods are part of a healthy food web, their numbers are controlled; they do not become the serious pests they can be in a garden where the use of chemicals and other damaging practices has thrown the system out of balance.

Chapter 11
Reptiles, Mammals, and Birds

W^E WON'T SPEND too much time on these larger animals. Many gardeners are plagued by them, but squirrels, mice, groundhogs, rabbits, chipmunks, voles, moles, prairie dogs, gophers, snakes, lizards—all burrow and travel in the soil, mixing, moving, and depositing organic matter and providing pathways and reservoirs for water and air. Still, most gardeners would just as soon never see one in their gardens either out of fear (reptiles) or hatred (groundhogs, rabbits, moles—you name it, if it burrows. And did we mention browsing moose and deer?).

The role these larger animals play in a vegetable garden is very different from the role they play in other parts of the yard. But wherever they roam, their role is important and entirely underpinned by microarthropods and microorganisms, which far outnumber them in any soil food web. The dung of all reptiles, mammals, and birds serves as a food source for other members of the food web community, which recycles it into nutrients. They also carry microbes on and in their bodies and feet from one location to another, and at death their carcasses are decayed by soil life.

The activities of larger animals are more easily observed and thus better known than those of other food web members; but like all forms of life, their

Chipmunks are always busy, and their activities affect the soil food web. Photograph by Paul Bolstad, University of Minnesota, www.forestryimages.org.

numbers depend on the habitat and foods they require. The presence of birds, in particular, indicates that larger arthropods, worms, and larvae are about, so seeing birds hopping across your lawn or even in your garden should give some comfort: a food web is in place and at work. Of course, the same can be said of moles burrowing about the lawn in search of Japanese beetle larvae. You may not want to have moles tunneling in your lawn, but since you know how the soil food web works, you at least know there is a food source somewhere that is supporting the mole population. This should inspire you to do something about the moles without resorting to chemicals and poisons.

Where do we humans fit into the soil food web? We have a huge impact on it, and very often not a positive one. Most gardeners have never heard of soil food web systems, even though they exist everywhere, and have no inkling of the role of microbes and arthropods play in them. And, of course, the gardener hardly ever knows when enough is enough and almost always tips the delicate balance a soil food web maintains.

Rototilling; spraying with herbicides, pesticides, fungicides, and miticides; compacting soil; removing organic material from lawns and under trees—all these human practices affect the soil food webs in your yard and gardens. Once a niche is destroyed, the soil food web starts to work imperfectly. Once a member of a niche is gone, the same thing happens. In both instances, the gardener must step in to fill the gap, or the system completely fails. Rather than working against nature, the gardener had better cooperate with it; and this, as we shall see, does not require a lot of hard labor—not if the gardener understands and teams up with the soil food web, letting its members do the work.

Robins are great microbial "taxicabs."
Photograph by Terry Spivey, USDA Forest Service, www.forestryimages.org.

Part 2

Applying Soil Food Web Science to Yard and Garden Care

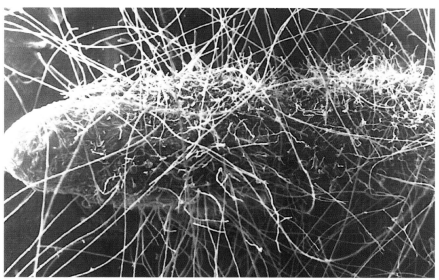

Mycorrhizal fungi, extending from a root—and increasing the plant's ability to obtain nutrients and water. Courtesy Mycorrhizal Applications, www.mycorrhizae.com.

Chapter 12
How the Soil Food Web
Applies to Gardening

Y OU NOW HAVE an appreciation of the many benefits a healthy, functioning soil food web provides you as a gardener. Of course, what's good for cut flower field trials in California will be different from what's ideal for row crops in Georgia, but no matter your climate, whatever your soil type—things will only improve when you put all those fungi, bacteria, protozoa, nematodes, arthropods, and other members of the soil food web to work for you, 24 hours a day, 7 days a week, to make yours a better yard and garden and you a better gardener.

First, a fully active soil food web will have better nutrient retention in its soils. The bodies of all its members hold (immobilize) materials that will eventually be broken down into plant nutrients. Every time a fungus or bacterium is consumed and digested by a protozoan or nematode, nutrients are left behind in plant-available form. And since plants attract fungi and bacteria to their rhizospheres, the nutrients they provide are in the right location to be easily absorbed.

Next, a healthy soil food web results in improved soil structure, starting with the efforts of bacteria that produce slime that binds the tiny individual soil particles into larger aggregates. Fungal hyphae, worms, insects and their larvae, and even small mammals travel through soil creating tunnels, big and small. This results in soils that have the right porosity, resulting in water retention and drainage as well as aeration, all necessary for healthy plant growth.

Soil food webs provide defenses against disease and those whose population growth and habits could throw the web out of kilter. Some members of the soil food web act like police, hunting down and capturing bad guys. Others act like doctors, dispensing vitamins and hormones. Fungi and bacteria serve as barriers around plants, blocking the entry of herbivores intent on getting to the plant roots, stems, or leaves; they also compete for the nutrients, space, and even the oxygen the bad guys need to survive.

Finally, soil food web organisms influence soil pH where it counts, right in the rhizosphere, which determines what kind of nitrogen is prevalent, nitrate or ammonium. A plant that attracts and receives its preferred form of nitrogen will perform optimally. Soil microbiology can even take care of pollutants, which is what lawn and garden chemicals really are, not to mention the pollutants in the air and in some instances, the water. In a healthy soil food web, there is something in the soil that eats almost anything you can find in the soil, including lots of the stuff man deposits there, purposely or inadvertently.

New rules

We have developed nineteen very simple rules to guide the gardener in using the soil food web (see the appendix for a recap of the whole list). Rule #1: some plants prefer soils dominated by fungi; others prefer soils dominated by bacteria. Plants need nitrogen to produce amino acids; it is crucial to plant growth and survival. This is why inorganic, soluble nitrogen fertilizers do a great job growing plants even while they are detrimental to the food webs. In water solution, these nitrates (NO_3^-) are readily available to plant roots, which pretty much act like sponges. As anions, they go into water solution instead of attaching themselves to humus or clays as positively charged cations would.

Two forms of nitrogen are available to plants when there is a healthy soil food web, nitrates and ammonium (NH_4); and—as in most things in life when there is a choice—some plants prefer their nitrogen as nitrates while others prefer ammonium.

When nematodes and protozoa consume fungi and bacteria, nitrogen is released in ammonium form in the waste stream. Ammonium is quickly oxidized or converted to nitrates by nitrogen-fixing bacteria when they are present in sufficient numbers in the soil. This is almost always the case when the soils are dominated by bacteria as compared to fungi because the slime produced by soil bacteria has a pH above 7, the right environment for nitrifying bacteria. In bacterially dominated soils, nitrifying bacteria generally thrive.

Fungi foster lower pH numbers because they produce organic acids to decay organic matter for nutrients. If there are enough fungal acids to offset the bacterial slimes, the soil's pH drops below 7, making the environment acidic and therefore more and more unsuitable for most nitrifying bacteria. More ammonium remains ammonium.

As a gardener you must appreciate that the plants in your backyard are not exceptions to Rule #1. The soluble nitrogen fertilizers you use not only suck the life out of the microbes in the soil food web, but they may not even be the best type of fertilizer for the plants you seek to grow. Usually, plants can survive utilizing even the less preferred form; however, most plants do better with one form of nitrogen over the other.

Who wants what?

The answer to what any given plant prefers is found in the next two soil food web gardening rules. Rule #2 holds that most vegetables, annuals, and grasses prefer their nitrogen in nitrate form and do best in bacterially dominated soils. Rule #3 points out that most trees, shrubs, and perennials prefer their nitrogen in ammonium form and do best in fungally dominated soils.

These two general rules take the guesswork out of what could have been one of the most difficult things about starting to garden with the soil food web. The rules make it easy to figure out what likes what, but once you understand what is behind them, you will appreciate them even more.

Early succession communities are bacterially dominated. As more and more organic litter accumulates in the waste products from these organisms and the plant life they support, fungal spores finally have enough nutrients at hand to germinate. With a place to take hold and the resources to support themselves, the resulting fungi thrive.

Many other factors are involved, but to stick to what concerns us: as plant life and the soil food web become more varied, fungal numbers increase and more short-lived plants like annuals give way to more permanent, perennial grassland plants. More organic matter is produced, providing food for ever-increasing fungal populations. Shrubs move in, followed by soft hardwoods, expanding saplings, mature hardwoods, and finally the kinds of conifers you find in old growth forests. All the while, fungal biomass grows in proportion to bacteria, which cannot possibly compete because they are limited to digesting

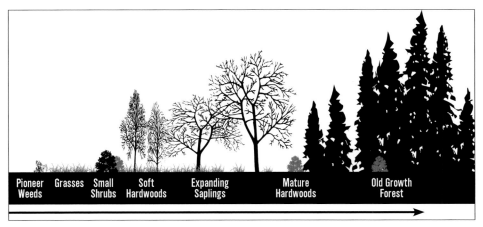

Plant succession, from the weeds on bare soil to old growth forest. Courtesy Tom Hoffman Graphic Design.

simple sugars and other carbohydrates—which are in limited supply given the ever-increasing mass of more complicated plants full of lignin and cellulose.

Moving from the beach, so to speak, to grasslands to old growth conifers, fungal dominance increases in the soil each step of the way. Part of this increase is explained by the tentative nature of early plant life. It is hard to form mycorrhizal relationships with plant roots when the plant dies after only a short period of time. You might as well live on your own as there is no advantage to a partnership.

It appears there is almost the same number of bacteria—100 million to 1 billion—in a teaspoon of garden soil, prairie soil, or forest soil. The difference in dominance of fungi over bacteria generally has everything to do with the increase in fungal biomass, not a decrease in bacterial biomass. Plants that would normally grow near the "beach side" of the continuum prefer bacterially dominated garden or yard soils and those that grow toward the other end, the old-growth-forest side, do best in fungally dominated soils. The transition occurs in the prairie plants, which like a balance between the two. This is analogous to your lawn grass, incidentally.

Another way to figure out what kind of nitrogen a given plant will prefer is to consider how long it lives. If it is only going to be in the ground for a season, as are vegetables and annuals, then you know the preferred form of nitrogen is nitrate. Anything in the ground for a year or more is usually going to prefer greater amounts of ammonium. This makes sense as well. Remember, bacteria numbers stay pretty much the same in all growing environments: it is the increase in fungal biomass that changes the ratio. Fungi are very fragile organisms and take time to grow. If they are mycorrhizal fungi, which many soil fungi are, then they have to have a live root with which to associate. The longer that root is alive, the longer the mycorrhizal fungi will be, so to speak. (It might not be longer, but rather may have more branches.) And finally, the litter from plants that live a season or so generally doesn't have the lignin and cellulose that are good fungal food sources. It is full of cellulose almost exclusively, which bacteria like. Bacteria reign.

	GARDEN	PRAIRIE	FOREST
BACTERIA	100 million to 1 billion	same	same
FUNGI	several yards	10s to 100s yards	1 to 40 miles (in conifers)
PROTOZOA	1000s	1000s	100,000s

Microbial populations (counts of bacteria and protozoa; lengths of fungal hyphae) in a teaspoon of various soils. Courtesy Tom Hoffman Graphic Design.

Fungal to bacterial biomass

For some of the specific garden plants you might encounter, the preferred fungal to bacterial biomass (F:B ratio) has been observed and measured. To match these, you can increase fungi by providing fungal foods or foster bacteria by providing bacterial foods; subsequent chapters will explain just how (or see the appendix for a summary of specifics on ways to accomplish this).

If you are a vegetable gardener, you need to aim for a biomass that has slightly more bacteria than fungi. More specifically, carrots, lettuce, broccoli, and cole crops prefer an F:B of 0.3:1 to 0.8:1; tomatoes, corn, wheat go for an F:B of 0.8:1 to 1:1. Lawns prefer an F:B ratio of 0.5:1 to 1:1. Agricultural testing labs will test your soil and provide you with an F:B ratio.

Trees require a higher F:B ratio. Forest soils, in which many of our landscape trees and shrubs originated, have a biomass of fungi over 100 times the biomass of bacteria. Conifers require the most fungally dominated soils with an F:B of 50:1 to 1000:1. Maples, oaks, and poplars require fewer fungi, an F:B of 10:1 to 100:1. Orchard specimens do best in soils with an F:B of 10:1 to 50:1; and some trees (alder, beech, aspen, cottonwood, and others that originate from riparian ecosystems) actually do best in bacterially dominated soils when they are young and fungally dominated soils (an F:B of 5:1 to 100:1) when mature.

Those of you who enjoy flowers will want to know that most annuals prefer bacterially dominated soils, while most perennials prefer fungally dominated soils. Again, the length of time a plant lives influences the rules.

Shrubs generally prefer a higher fungal dominance than perennials (they are long-lived, so this follows our rule). Those native to conifer forests as opposed to deciduous forests require high F:B ratios; rhododendrons, for example, require very strong fungal dominance, while a cotoneaster or lilac requires less.

There are more rules to come—but don't forget Rules 2 and 3, because the management of nitrogen is fundamental to success in the yard and garden.

Chapter 13

What Do Your Soil Food Webs Look Like?

I

F YOU ARE GOING TO USE the science to which you have just been exposed, the very first thing you need to know is the current state of the soil food webs in your yard. Once you have a base established, you can figure out what needs to be done in order to end up with the best possible soil food web for whatever you happen to be growing.

Take a census

Did we say soil food webs, plural? At this point in the book, this should not be a big surprise. Different plants produce different exudates that attract different bacteria and fungi. These in turn attract varying predator organisms. So, as you would expect, the soil life around the roots of trees on one side of your house is going to be completely different from what surrounds the roots of your vegetables, which is different from the soil life that supports your lawn and possibly even the same trees on the other side of your property.

Areas that have been exposed in the past to commercial fertilizer will have less soil life than areas kept in a natural state. Parts of your yard that have been heavily compacted or frequently rototilled will have fewer fungi and worms than the areas you've left alone. You may have an orchard or a foundation planting of conifers. It is important to figure out what life makes up the various soil food webs in your yard. In order to do this, you must go hunting in the soil and take a census of what is there.

The images in this book have forewarned you: you may find things in your soil that, upon closer examination, will scare the daylights out of you. (In general we advise against putting anything under an electron microscope. At that level, all life has teeth!) The point is, when you get a good look at some of the microarthropods present in soil, you may never want to put your hands in the soil again. Sometimes ignorance really is bliss; however, in this instance a little knowledge is not going to hurt you and will actually help you be a better gardener. Just remember, you put your hands in the soil before you knew what was there and never got hurt.

You will want to repeat the following procedures with soils from each of your gardens and lawn areas, and even around specific trees and shrubs. We have done this dozens of times in our own yards, and what we find never fails to astonish us.

Find the bigger animals first

Start by digging a hole in the soil at issue, about 12 inches (30 centimeters) square. Use a spade or trowel—it doesn't matter, and measurements don't have to be exact. Put all the soil you dig up onto a tarp or in a box so you can then sift through it, looking for the bigger animals you might find in the soil: worms, beetles, insect larvae—any living organism you can see with the naked eye and pick up without having to resort to tweezers. Keep track of what you are finding.

None of us are trained at identifying all the organisms in our soils, and frankly the variety of them is so great as to be beyond the scope of this book. Do your best in making identifications. Seek help from others. In time you will become sufficiently proficient for the purpose. This is new stuff, and just being exposed to it will make the learning experience easier. It didn't take us very long, and it won't take you long to become familiar with soil food web organisms.

If you find worms or their castings, it is a good sign. Remember that worms serve as foods for small mammals and eat bacteria, fungi, protozoa, and the occasional nematode. If worms are in your sampling, most probably a whole set of soil food web organisms are busy at work in that soil—and it's probably good, rich, organic, nicely textured soil, at that. Similarly, the presence of millipedes and centipedes, beetles, spiders, springtails—even a few slugs and snails—indicates a healthy soil food web. If you find these, you have a good head start. You are already teaming with microbes, not to mention macro-arthropods and worms.

To make sure you are really capturing what is in your soils, however, you need to set soil traps. Many soil food web organisms roam at or above the surface of the soil for all or part of the day. To count as many of these as possible,

	GARDEN	PRAIRIE	FOREST
ARTHROPODS	<100	500–2000	10,000–25,000
EARTHWORMS	5–30	10–50	10–50

Number of visible organisms in a square foot of various soils. Courtesy Tom Hoffman Graphic Design.

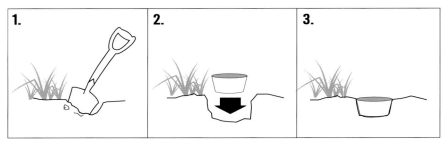

A simple trap will allow you to take a census of the larger animals in your soil. Courtesy Tom Hoffman Graphic Design.

you will need to bury a quart-sized container in the soil so that its lip is at ground level. If you are in rain country, prop up some kind of cover (an open umbrella will do) to keep water from getting into the trap. Next, pour in half an inch of animal-safe automotive antifreeze or toss in a mothball or two, and leave it alone for several days to a week. Make as many traps as areas you are testing.

Unsuspecting gastropods and macroarthropods will fall in these simple traps to be counted later. Give traps a visual check every now and then to see what has been captured. If you have kids or pets, skip the antifreeze and use mothballs at your discretion. Both are used to kill entrants to the trap (so they don't eat each other and mess up the census); they are not attractants, so they are not absolutely necessary. By the end of the week, you should find a few of the larger arthropods such as beetles, millipedes, and centipedes. You might also find some slugs and even a worm or two.

Take inventory of all your traps. Traps empty? This means you need to do a lot of work to restore the soil food webs in your yard. If few of the larger participants in soil food webs are present, some link or links before them on the food chains that make up the web are missing.

Count the smaller organisms

Surveying the microarthropods requires a different kind of trap—a Berlese funnel, named after Giovanni Berlese (1863–1927), the scientist who invented it.

You can easily make your own Berlese funnel. First cut the bottom off a liter-sized plastic bottle, the kind soda or juice comes in. Turn the bottle so the drinking end is facing down (this is the funnel). Next, place a 2-inch-square of window screening with openings approximately $1/16$ to $1/8$ inch (1.5 to 3 millimeters) inside the bottle so it settles in the neck. Nothing larger than the openings of the screen will settle through.

Next, set the mouth of the bottle into a quart-sized container. The container has two purposes. The first is to act as a repository to collect the organisms that fall into it through the screen and down the neck of the funnel. The second is to hold the funnel and give it stability. It is, after all, an upside-down soda bottle and doesn't balance very well. We use large, recycled yogurt or cottage cheese containers because they are just the right size to hold such a bottle and are really easy to come by.

The next step is to fill the funnel with soil and duff, the organic debris that is on the top few inches of many soils. Start with a particular garden or your lawn and sample down to about 8 inches (20 centimeters).

If you want to do things a bit more scientifically, pour a bit of antifreeze or ethyl alcohol into the holding container so it just covers the bottom. Either of these will kill all the organisms that fall in so they don't eat each other before you get to observe your catch. You can skip this step with no fear of the critters leaving the container; the plastic is too slippery. A few organisms will be lost to cannibalism, as the feeding frenzy that goes on in the soil continues in the container; this can be a morbidly fascinating show.

Next, apply heat. This gets the life in your mixture to move from the soil (where it is perfectly comfortable) down into the container. Suspending a 40- to 60-watt lightbulb over the open end of the funnel (or placing it under an existing light source of similar wattage) accomplishes this. The top of the funnel should be about 6 inches (15 centimeters) from the source of the heat. Be careful: you can have the best soil food web going, but if you burn the house down by overheating the materials in your Berlese funnel, your spouse is not going to be happy no matter how well the garden turns out.

Turn on the bulb and leave the Berlese funnel undisturbed for at least three days. Its light and heat will drive the soil organisms down through the screen into the wading pool in the container. Some folks put a few mothballs on the top of the soil instead of using the heat from a lightbulb with the same results: a mini-stampede of microarthropods and other organisms into your observatory. You can peek as often as you want, but don't stop the process for a minimum of three days (a week is best) if you expect to get all the life you can into the container.

Now it's time to count your catch. Your best bet is to look at the container's contents with a magnifying glass or a MacroScope, a monocular that allows you to stand back at arm's length but view the trapped microarthropods and occasional gastropod as if they were only a couple of inches or so from your eyeball.

What astonished us (and frankly, despite all the research we have done for this book, remains just as amazing) the first time we did this were the number of living things we saw—mites, the larval stages of a half-dozen animals, tiny beetles, springtails, and more. We simply had never seen most of these before. One might expect this to be true when the subject matter involves microbes, but as lifelong gardeners (who have lived a fairly long life and spent an awful lot of time digging around in the soil), we thought we knew what was living there. How wrong this proved to be. We are pretty sure you will be equally amazed and feel the same kind of astonishment.

Since populations of micro- and macroarthropods are different in different areas and since it is important to have at least a general idea of what is normally in your soil food web, you may need to contact your local cooperative extension or other governmental agricultural agent for information on what you have collected. You can also contact the nearest university. And many resources are available on the Internet.

Admittedly, these censuses are not perfect. You are catching what happens to be going by at the time the traps are set, but it is a snapshot of at least the mobile members of the soil food web. A diverse and numerous population of microarthropods in the catch basin of a Berlese funnel is a good sign things are on track and suggests good populations of microorganisms too are present. Similarly, a lack of diversity and numbers should cause some concern; you'll need to do some work to return them.

"Counting" microorganisms

How do you assess the population of microorganisms, which are, after all, the major source of nutrient retention and cycling in the soil food web? The numbers of nematodes, protozoa, bacteria, and fungi will help tell you what nutrients are available to your plants and the ability of your soils to mineralize and immobilize these nutrients. If you know what is in the soil, you know what is missing—but when it comes to the microorganisms, we will be the first to admit that you will not be able to determine precisely what is in your soil, even with a powerful microscope. You will be able, however, to identify nematodes, some protozoa, and algae, and at least see (but not identify) bacteria. Exact measuring is best left to the professionals.

But first, let's make some inferences. If you found lots of earthworms in your samplings, there is every likelihood that your soils contain good bacterial and protozoan populations and more, because these are what worms generally eat. And in some soils and mulches, you can see evidence of fungi, either mycelia (where there is decaying organic matter) or fruiting bodies in the form of mushrooms. If you already care for your property without the use of pesticides, fungicides, and salt-based fertilizers, and you know that organic matter placed in your yard and gardens decays rapidly (within six months in a temperate zone), you have yet another strong indication that the base members of your soil food webs are at least fairly healthy.

You may not want to, but you can measure some nematode populations to a certain degree. First, take a Berlese or kitchen funnel and put a hand's-length section of surgical tubing over its smaller end; clamp the tubing shut with a large paper clamp. Next, collect a few handfuls of soil and mix them with dechlorinated water, forming a thick but soupy mud. Fill the funnel half full with the mud, and then pour more water in, so it covers the mud and then some. The nematodes will sink into the funnel's neck. After 24 hours, quickly open and close the clamp, and examine the concentrate you've just released with the best magnification you have. A microscope and a few drops on a slide could produce a great show.

But again, to get a really accurate assessment of microbial populations requires the training and the sophisticated lab equipment of a professional. Traditional soil tests determine elemental deficiencies in the soil (NPK tests) and measure the soil's pH and CEC. These are useful, but for soil food web purposes, quantifying the amount of fungi and bacteria, especially, is critical.

It is pretty easy to get soils tested for nematodes by almost any lab that does agricultural soil tests; and protozoa can be seen in relatively inexpensive microscopes. If you have good numbers of beneficial nematodes and low-to-no bad guys, you know you have good nutrient-cycling capabilities. The same is true if you have lots of protozoa. But what you also want to know, from any lab that tests your soil for biology, are biomass numbers. How much fungal biomass is there in the soil? How much bacterial biomass is there? This is where the nutrients are stored—in the bodies of fungi and bacteria. This information will determine which type of organisms dominate in your soils and in what ratio they exist.

More and more agricultural testing labs are recognizing the value of testing soils for their microbiology. You should be able to find a lab to take care of your needs (an analysis of a compost sample, courtesy Soil Foodweb, Inc., www.soilfoodweb.com, concludes this chapter). Armed with the results of your own visual surveys and such microbiological lab tests, you will know what is active in your soil and, by implication, what isn't there. Next, you need to learn what you can do to maintain and support existing members of the community. But rest assured: whoever is missing can be activated by soil food web gardening techniques.

Soil Foodweb, Inc
728 SW Wake Robin Avenue,
Corvallis, OR 97333 USA
Phone: (541) 752-5066
Fax: (541) 752-5142
e-mail: info@soilfoodweb.com

Organism Biomass Data

Sample #	Unique ID	Dry Weight of 1 gram Fresh Material	Active Bacterial Biomass (µg/g)	Total Bacterial Biomass (µg/g)	Active Fungal Biomass (µg/g)	Total Fungal Biomass (µg/g)	Hyphal Diameter (µm)	Protozoa Numbers/g Flagellates	Amoebae	Ciliates	Total Nematode Numbers #/g
363	NW Vermi	**0.31**	188	4,002	46.0	4,928	2.75	1,136,894	146,682	1,831	48.1
364	KIS-Thermal	**0.30**	468	2,193	32.7	5,959	3.00	469,291	19,478	1,557	67.2
Bold Means Low		Both Too wet: allow material to dry out a bit, to prevent anaerobic conditions.	Both Excellent.	Both Excellent.	Both Excellent.	Both Excellent.	Community of disease-suppressive fungi present in both.	Excellent protozoan numbers. This material will provide a good inoculum of protozoa when applied to the soil. High ciliate numbers indicate good structure in the compost. The aggregates may be anaerobic on the inside, but as the anaerobic materials diffuse out of the aggregates, they encounter aerobic conditions, as indicated by the high numbers of flagellates and amoebae. This indicates a broad diversity of microsites, and therefore excellent diversity of bacteria and fungi.			Good numbers and diversity. Possible switchers present. Need to maintain adequate fungi to protect plants.
Desired Range		0.45 - 0.85	15 - 25	100 - 3000	15 - 25	100 - 300	(A)	10000 +	10000 +	50 - 100	20 - 30

Immature compost can have activity ranging from 10 to 100%. Mature compost should have activity between 2 to 10%.

Fungal activity and biomass depends greatly on the plant being grown. Desired range given here is for a 1:1 compost.

A - Hyphal diameter of 2.0 indicates mostly actinomycete hyphae, 2.5 indicates community is mainly ascomycete, typical soil fungi for grasslands, diameters of 3.0 or higher indicate community is dominated by highly beneficial fungi, a Basidiomycete community.

Season, moisture, soil and organic matter must be considered in determining optimal foodweb structure.

If sample information, such as pesticide, fertilizer tillage, irrigation are not included on the submission form, sender's locale is used.

One report is sent to the mailing address on the submission form.

All submissions receive free 15 minute consultation, call 1-541-752-5066

00363: Mature compost from NA. Smell: Mild.
  For use in tea brewing.
00364, Mature compost, mild odor

Sample	Assay Activities	Notes
363		Actinos present
363	T.F.	Good diveristy and hyphae diameter ranging from 1.5 to 8.0
364	T.F.	Great diversity with diameter ranging from 1.5 to 20 and mostly 3 and lots of long hyphae

Organism Ratios

Sample #	Unique ID	Total Fungal To Total Bacterial Biomass	Active to Total Fungal Biomass	Active to Total Bacterial Biomass	Active Fungal to Active Bacterial Biomass	Plant Available N Supply from Predators (lbs/acre)	Root-Feeding Nematode Presence
363	NW Vermi	1.23	0.01	0.05	0.24	300+, but N loss	None detected
364	KIS-Thermal	2.72	0.01	0.21	0.07	300+, but N loss	None detected
		Fungal dominated compost, suitable for variety of plant applications.	Fungal component is mature.	NW vermi/ bacterial component is mature. K is thermal Not mature. Wait to apply this material until activity drops below 10%. Material is currently suitable for making tea.	Compost will become more bacterial with time,	Excellent nutrient cycling. N loss results from anaerobic conditions, as indicated by high ciliate numbers.	Possible switchers present. Need beneficial fungi and nematodes to combat these pest conditions.
Desired Range		*(1)	*(2)	*(2)	*(3)	*(4)	*(5)

(1) For the following plants, Grass:0.5-1.5; Berries, Shrubs, grape: 2-5; Deciduous Trees: 5-10; Conifer: 10-100.

(2) Active organisms in mature compost should be below 0.10. Compost is not mature, i.e., not stable, if greater than 0.10.

(3) For annuals, ratio should be 1 or less, for perennials, ratio should be 2 or greater.

(4) Based on release of N from protozoan and nematode consumption of bacteria and fungi. Often protozoa and nematodes compete for food resources. When one is high, the other may be low. Also, if predator numbers are high, the prey may have low numbers

(5) Identification to genus.

Nematodes per Gram of Compost

	363	364
Bacterial Feeders		
Butlerius	4.86	1.04
Cuticularia	7.42	14.62
Eucephalobus		0.35
Mononchoides	0.77	
Plectus		1.04
Rhabditidae	1.53	1.04
Rhabdolaimus		0.35
Fungal Feeders		
Aporcelaimus		0.35
Mesodorylaimus		0.35
Fungal/Root Feeders		
Aphelenchus	0.26	
Ditylenchus	0.26	0.70

Chapter 14

Tools for Restoration and Maintenance

NOW THAT YOU HAVE an idea of what populates your soils, it is time to take whatever action is necessary to ensure your soil food webs give your plants what they need in the way of nutrients and protection.

Compost, mulch, and compost tea

This is when you begin teaming with microbes and become a soil food web gardener. With most soils, your first aim will be to restore a diverse and whole soil food web. As beneficial organisms return, you will see a difference not only in your soils but in your plants as well. Some areas (lawns and beds of annuals, for example) respond very quickly; other spots will have soil food webs that take longer to establish or alter. Much of your yard's response will have to do with previous practices. If in the past you saturated your yard with commercial pesticides, herbicides, fungicides, or salt-based chemical fertilizers, you may have to completely reestablish soil food webs; this may take a year or more. Gardeners who have been "organic" usually need only to tweak their established food webs, employing some new practices and intensifying others.

It's simple. Compost, mulch, and compost tea are the soil food web gardener's tools, and it takes only three strategies to restore the soil food web using them: applying the proper kind of compost; mulching the right way, with the right kinds of organic matter; and applying actively aerated compost teas (AACTs). Once established, soil food webs can be maintained with the same strategies, either alone or in combination. Employed properly, these management tools will replace conventional fertilizing with chemicals. These tools feed the microbes that feed the plants. If you keep the microbes happy, healthy, and diverse, you will have excellent results.

Compost has been used to support soil food web organisms long before anyone knew they existed. It is a proven, effective growing medium. Compost can inoculate an area with microbes to support a soil food web. Properly made compost contains the entire complement of soil food web microorganisms: fungi and bacteria, protozoa and nematodes. It is also full of organic matter, which provides living space and nutrients for the gang of microbes a compost pile contains. Finished compost never smells bad, which would be a sure sign of anaerobic microbes doing their thing. It should smell earthy and fresh, and it always has a rich, dark, coffee color. The only caveat is that in modern times, one has to know what was used to make the compost, as many of the chemicals we seek to avoid do not break down quickly enough in compost.

Organic mulches, too, are an effective soil food web gardening tool. By organic we mean natural material, full of carbon and nitrogen—namely, leaves, grass clippings, and wood chips. These provide the proper environment for the soil community's organisms and plenty of organic foods for them to live on. After all, these are what make up the compost pile. Mulch is a form of cold compost: it doesn't heat up like a compost pile, but it will decay, over a longer period of time. By providing different kinds of organic matter as mulch, you can establish or supplement different members of the soil food web, ones that will provide more of the type of nitrogen preferred by the plants grown in the area.

Actively aerated compost tea is a liquid easily extracted from compost. A properly made AACT contains the same set of microorganisms as the compost from which it derives. The term "actively aerated compost tea" is used to distinguish these modern compost teas from old-fashioned teas like the ones your parents and grandparents may have made by soaking a bag of compost or manure in water for a few weeks. AACTs are prepared by pumping air into a mixture of compost, dechlorinated water, and microbial nutrients. Unlike old-fashioned teas, which went anaerobic, AACTs remain aerobic—and the aerobic microorganisms are the beneficial ones. The energy from the air bubbling through the mixture strips the microbes out of the compost and into the tea. Here they grow and multiply, forming a stew of beneficial food web microbes that can be applied to soil.

Aerated compost teas are easier to make and much easier to apply than compost and have a higher concentration of microbes, so you don't need nearly as much tea as you would regular compost to inoculate an area. These teas can also be sprayed on leaf surfaces, where compost will not stick. Here the beneficial microbes in the tea outcompete pathogens for food and space.

More work now, much less later

Using compost, mulch, and compost tea properly will greatly reduce the amount of work it takes to maintain your yard and gardens. There is a bit of work involved in making the conversion from chemicals to microbes, but ultimately, once you gear up and make the necessary changes, there will be less to do. The microbes will be working for you. You will need to water less because the food web animals will have improved your soil's water- and air-holding capacity. You won't need to fertilize because there will be proper microbial cycling of nutrients in the soil. And you will be able to ensure your plants are getting the kind of nitrogen they prefer.

You will have fewer plant health problems and some effective, easy-to-use tools to make things better if things do go wrong. And if all this doesn't save you time and effort, not having to rototill or turn your garden soils—ever again—surely will. Best of all, there are no dangerous chemicals; nothing leaches into the water table. When you team with microbes, there is no small print to read—and no health problems for you, your family, or your pets.

You have now heard, briefly, what the main soil food web tools are; each deserves and will get its own chapter. Once you start applying all the rules using these three tools, we are quite sure there will be no looking back.

Chapter 15
Compost

COMPOST is a whole universe of diverse soil food web organisms. Never mind the huge numbers in good, fertile garden soil: the numbers of organisms per teaspoon in compost, especially the microbial populations, are simply too large to fully comprehend: up to a billion bacteria, 400 to 900 feet (150 to 300 meters) of fungal hyphae, 10,000 to 50,000 protozoa, and 30 to 300 nematodes. In addition to extremely high microbial numbers, compost contains all manner of microarthropods and sometimes worms. It teems with life.

Rule #4 (compost can be used to inoculate beneficial microbes and life into soils around your yard and introduce, maintain, or alter the soil food web in a particular area) establishes the use of compost as a major soil food web tool. Rule #5 elaborates on this: adding compost and its soil food web to the surface of the soil will inoculate the soil with the same soil food web. The organisms in the compost you apply to your gardens, trees, shrubs, and perennials will spread life as far as they can. It is microbial manifest destiny. But you can best satisfy a plant's nutrient needs by adding compost with the right microbial domination.

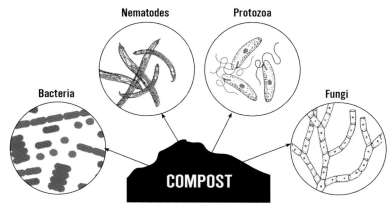

Compost contains the key soil food web organisms that hold as well as cycle plant nutrients. Courtesy Tom Hoffman Graphic Design.

Not all composts are the same

Most gardeners don't give compost much thought. They make or buy it, and they apply it—it is all the same. There is more than one kind of compost, however, which is something that amazes many veteran compost makers. We, too, thought that all compost, no matter what went into it, had the same biology and pH in the end. But surely, upon reflection—and especially after you know something about the soil food web organisms that make up compost—the idea that the end product is always the same doesn't make any sense. As with almost every other system, what goes into it does have something to do with what comes out at the end.

The fact of the matter is that by using just a bit of soil food web science, you can make either compost that is dominated by fungi or compost that is domi-

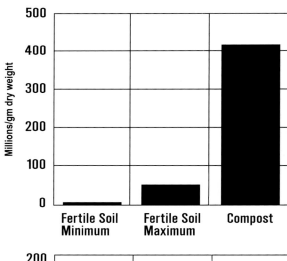

Bacteria populations in fertile soil and compost. Courtesy Tom Hoffman Graphic Design.

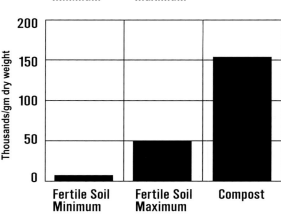

Fungi populations in fertile soil and compost. Courtesy Tom Hoffman Graphic Design.

nated by bacteria. It all depends on what you put into the compost pile or bin to start. And because some plants prefer their nitrogen in ammonium form and some in nitrate form (see Rules 2 and 3), making compost that fosters the production of one of these forms of nitrogen over the other makes real sense.

How to make compost

Farmers have been using compost to improve their soils at least since the time of the early Romans. It was only in the last century that compost took a backseat to chemicals when it came to growing things; before that, if you worked on a farm or in a garden, you routinely used compost and manure to increase fertility. This all changed when internal combustion engines replaced the horse, and fewer and fewer homes, particularly in urban settings, featured chickens, cows, pigs, and other livestock. Agriculture and horticulture required chemicals because there was a dearth of manures and thus compost.

Making and using compost has made a strong comeback among home gardeners and has even become politically correct: composting conserves valuable landfill space by recycling at least some of our household wastes. Dozens of compost bins are commercially available, and a like number of books can tell you how to make compost, in myriad ways. At the heart of every composting system, however, are the soil microorganisms, the members of the compost's food web. They are the ones that make compost, no matter what method is employed. Their metabolic activity creates the heat and by-products that make the composting process work.

This is a chapter, not a book, on composting. What we will describe here is just a bit of the science behind composting and a few basic procedures for making compost at home. Once you have made a few batches, you can experiment and create a system that best fits your plants' needs and your climate, space availability, and even spousal demands. Besides the necessary soil microbes, composting requires heat, water, air, and organic materials with the right amounts of carbon and nitrogen. All are mixed in the proper ratio.

Organic materials are easy to come by: grass clippings, autumn leaves, wood chips, straw, sawdust, branches, and virtually all kitchen scraps (except meats and fats). Human and pet feces should not be composted because of the possibility that disease organisms might survive even the high heat of the compost process; for the same reason, we personally discourage the timeworn practice of using other manures in compost. Why take the risk when you don't know what kind of antibiotics and other drugs were used to feed the animals? Who wants to be worried about *E. coli*?

Bacteria, fungi, and other microbes seek mostly carbon from the organic matter in the compost pile, just as they do in the soil. This fuels their metabolism. The microbes also need nitrogen to make the enzymes used in the decay process and the proteins (including their chief component, amino acids) that are necessary to build structure and enzymes.

Moisture is necessary to provide the optimum environment for the microbes and to prevent them from dying or going into dormancy. You cannot have active bacteria, protozoa, or nematodes without the water necessary for their transport and other life functions.

Air is needed because the beneficial soil organisms that break down carboniferous and nitrogenous materials are aerobic. They breathe air; they require oxygen. It is true that anaerobic conditions can develop in a compost pile and decay will occur under these conditions as well; however, so will the production of things detrimental to plants, such as alcohols, of which as little as one part per million will kill plant cells. Obviously, then, it is important to keep compost piles aerobic, which is why compost piles are turned and opened up, bringing air into the system.

Finally, the heat required for composting does not come from the sun but rather from the soil life's metabolic activity, most of it from bacterial activity. As you will see, this heat is what creates an environment that increases populations and causes them to change in character at the appropriate time during the composting cycle.

Mix these ingredients in the right proportions, and you will end up with a rich, crumbly, dark, coffee-colored, sweet-smelling humus-soil that also happens to be full of life. Though it could take as long as a year or more, it is possible to make good compost in as little as a few weeks. But no matter what method is employed, it is the microbes that do most of the work.

Mesophilic and thermophilic stages

Composting material goes through three distinct temperature phases. The first of these phases is the mesophilic. Mesophilic organisms thrive in moderate temperatures, between 68 and 104F (20 and 40C).

Even in this first stage, work begins on the straight, difficult-to-digest chains of cellulose, which are broken into smaller chains of glucose; bacteria are particularly adept at depolymerization, as this process is known. Meanwhile, brown rot fungi (basidiomycetes, "regular" mushrooms) and certain bacteria (*Bacillus* spp., *Heliospirillum* spp.) are active breaking down other difficult-to-digest material. These microbes produce endospores, spores that are resistant to chemicals and heat; this enables them to survive the next, hotter phase of composting, and they return when temperatures cool.

Larger soil organisms join the fungi and bacteria, breaking apart organic matter in the pile as they search for food, and microbial activity in the guts of some of these animals results in further chemical breakdown. All this metabolic activity creates heat, raising the temperature to 104F (40C). At this point, it becomes too hot for the continued activity of the mesophilic organisms, and those adapted to higher temperatures take over.

In case you are wondering how a compost pile heats up in the spring after a frozen winter, it is simple: some bacteria are psychrophilic, meaning they thrive at temperatures just above freezing though some of them—the really "cool" bacteria—can continue to operate at temperatures as low as 32F (0C). The metabolic activity of these cool-loving bacteria increases the temperature of the pile just enough to wake up the higher-temperature, mesophilic organisms so they can take over.

Organisms in the second stage of the composting cycle, the thermophilic phase, can withstand temperatures of 104 to 150F (40 to 65C) and over. During this period the complex carbohydrates are fully broken down. Some proteins are also decomposed. Hemicelluloses, more resistant structures, are decayed. Many more bacteria (*Arthrobacter* spp., *Pseudomonas* spp., *Streptomyces* and other actinomycetes) and fungi join in or begin to play more prominent roles. Their metabolic heat causes the temperature in the compost pile to continue to rise; these high temperatures also kill off pathogens that might be in the mix.

These first two stages take place very rapidly. A properly made compost pile should heat up to 135F (57C) in 24 to 72 hours; typically, if you have the right mix of carbon to nitrogen, the center of a pile will heat up to 135F (57C) in a day and 150F (65C) in three. If the pile is not heating up, then you need to turn it (that is, switch the inside and bottom materials in the pile with the outside and top materials) to add oxygen. If that doesn't work, add fresh, green material (as these are full of easy-to-digest sugars that will supply bacteria the food they need). Newspaper, fruit pulps, or commercial compost inoculums can also be added to help a pile heat up.

You have to monitor compost piles. It is advisable to keep a pile between 140F (60C) and 150F (65C) for at least a few days because at this thermophilic temperature, pathogenic microbes in the compost are killed. At 150F (65C), weed seeds are also destroyed. Never let a compost pile get over 155F (68C) as this will start to burn off carbon. To temporarily cool an overheated pile, turn it (yes, turning encourages both heating and cooling). Not only does this open

Turning a home compost pile. Photograph by Judith Hoersting.

the pile up to air, it ensures all the material in the pile gets treated. If turning doesn't do the cooling trick, add water or more brown materials, changing the ratio of green (easy-to-digest bacterial foods) to more fungal foods. Since bacteria are the primary heat-generating organisms, this will slow things.

There is nothing wrong with sticking your hand into the pile to gauge the heat. Or you can stick a long, gutter nail or metal rebar pipe into the pile; these transmit heat and will feel warm when things are going right. A thermometer is more precise, however; you can buy a soil thermometer designed for the purpose or use an oven thermometer.

Maturation stage

As the complex proteins and carbohydrates are broken down and begin to diminish, there is a reduction of metabolic activity and the temperature in the pile starts to decrease. The mesophilic organisms, whose specially protected spores enabled them to survive the higher heat stage, reassert themselves and replace the thermophilic organisms. The compost enters the final, maturation stage.

During the maturation stage, the decay of the most resistant plant component, lignin, is completed. The bonds holding the chains of alcohols in lignin together are extremely strong and structurally much more difficult to attack and break apart than almost anything else in the pile. The actinomycetes, the chain-like bacteria that resemble fungi, continue their attack on these really difficult-to-digest plant remnants; these are the same organisms that impart the earthy smell associated with good compost and soil, which comes from their decay of cellulose, lignin, chitin, and protein. The major fungal participants in this last stage, the basidiomycetes, are still at work.

Also during this maturation stage, physical decomposers continue to support the microbial team. Grazing by nematodes, springtails, centipedes, and others cause the populations of fungi and bacteria to increase; and as these microbial populations increase, so do their soil-binding activities. Lots of nematodes were killed by the heat of the thermophilic stage, but those that survive have lots and lots of bacteria and fungi to eat; as a group, they do well. Worms, too, work the organic matter in the pile, exposing it to bacteria and then coating particles with a mucus that binds them together into aggregates. Ants, snails, slugs, mites, spiders, rove beetles, and sow bugs can come into the pile and open up the organic matter as they forage, shredding it and making it easier for microbes to attack. The end result of all these organisms going about their day-to-day business is compost.

It is best to keep the compost pile between 104 and 131F (40 and 55C) after the initial thermophilic run-up to 150F (65C). Make sure that the outside of the pile gets turned into the center so all the material decays. If the pile drops below 104F (40C) before it is mature, consider adding some more green, high-in-nitrogen material. If it stays above 131F (55C), consider adding more brown, carbon-containing material. Of course, aerating a pile will always initially cool it down, and if you have the strength, repeated turning is the only control you need. Watering a pile down will also cool it, but this is a more drastic step.

The pile needs to remain moist throughout the process. Don't let it dry out, but don't let it become so saturated that there is no air supply in the pile. You may have to add water as you turn the pile, or cover it to keep rain from soaking it. If all goes well, and it usually does, "compost happens." After two or three turns, your pile should be compost. It is finished, or mature, when you cannot recognize what's in it.

C:N ratio and fungal vs. bacterial dominance

The ratio of carbon to nitrogen has to be right in order to make compost; the ideal C:N ratio for this purpose is somewhere around 25:1 to 30:1. If you have too much carbon, nitrogen is quickly used up and the decay process slows. If you have too much nitrogen, organisms snatch it up and then carbon is vented to the atmosphere or mixed with water and washed out of the pile. But at the ideal ratio, things go fast, and decay is complete.

Often gardeners divide available composting materials into two categories, brown and green. Aged, brown organic materials support fungi, while fresh, green organic materials support bacteria (Rule #6). Brown items—including autumn leaves, bark, wood chips, twigs, and branches—contain carbon; carbon provides members of the soil food web with energy for metabolism. Green items—such things as grass clippings, fresh-picked weeds, kitchen scraps—contain plenty of the easier-to-digest bacterial foods and are good nitrogen sources. The fresher the green item, the more nitrogen it will contribute to the pile. Nitrogen provides soil food web organisms with building blocks for proteins, which are used, among other things, to produce the digestive enzymes necessary in the decay process.

Not all organic wastes at hand have the ideal C:N ratio; sawdust, for example, is 500:1, and paper is 170:1. The two organic wastes you should have a good supply of are grass clippings (19:1) and tree leaves (40:1 to 80:1)—mixed together, these will give you near the proper ratio.

It is possible to manipulate compost materials so that the end product is highly fungal or highly bacterial, or a balance of the two: simply increase brown materials (to increase the amount of fungi) or green materials (to increase bacterial counts). A good mix of materials for a fungal recipe is 5 to 10% alfalfa meal, 45 to 50% fresh grass clippings, and 40 to 50% brown leaves or small wood chips. A suitable bacterial recipe would include 25% alfalfa meal, 50% green grass clippings, and 25% brown leaves or bark.

Again, the green materials that go into compost provide simple, easy-to-use sugars and lots of nitrogen and are great for supporting bacteria. The brown materials in compost piles consist of difficult-to-digest lignin, cellulose, and tannin (and some nitrogen as well). Fungi prefer this kind of material and have the enzymes to break it down. Only then can bacteria attack it.

Other important factors

The bacteria in compost will tend to buffer pH around 7 to 7.5. Fungi in compost will tend to buffer the pH around 5.5 to 7, so you want some fungi in all your composts to prevent them from getting too alkaline. The more fungal material in your compost, the lower the pH, to a point.

Inorganic fertilizers, pesticides, herbicides, miticides, and fungicides kill off soil food web members and therefore have no role in composting. Materials that go into compost piles should be free of these chemicals. Chances are they will break down over time, but maybe not before the compost is spread; and why take risks with chemicals when you don't have to? In addition, since many of these chemicals are nonselective when it comes to microbes, they can interfere with the composting process itself by eliminating microbes that contribute to the heat and decay.

The size of the material put into a compost pile is also important. Too much fine, particulate matter, and the pile will compress and quickly go anaerobic. If the material is too big, there will be so much air diffusing through it that the pile will heat up too much. If the material is too large, it won't decompose properly or fast enough because the bacteria can't get into it quick enough to establish sufficient populations. There is a fine balance when it comes to size of materials put into a compost pile, and only experimentation will give you the understanding you need and, finally, the control you want.

Next, a compost pile requires a minimum amount of mass, approximately 3.5 feet square or round (1 cubic meter), in order for it to heat properly. You can make your piles bigger, but the increase in size creates more work, as the entire pile has to be aerated or turned at least a few times to keep it from going anaerobic. In our experience, a six-foot pile, wide and tall, is about as big as you will want without a lot of mechanical help turning and aerating it.

It is quite easy to make compost literally in a pile, dumping ingredients

A professional turns his compost to aerate it. Photograph by Ken Hammond, USDA-ARS.

right on the ground and mixing them. Some prefer a caged area to contain the material and to make turning easier. A single ring of fencing or chicken wire, three feet in diameter and four or five feet high works great. Using a wooden pallet or screen supported on concrete blocks at the bottom of the pile will allow air to circulate into the pile, making it much less work to maintain. Some composters swear by revolving bins for making compost: in go the organic materials and to aerate all you do is spin the drum a few times. Once you figure out how to keep the materials in the bin from getting too moist (a chronic problem with enclosed systems), these can be very effective. Again, you will need to experiment to suit your tastes and needs.

Whatever your setup, you will need to keep an eye on the pile's moisture. Place materials in layers of 4 to 6 inches (10 to 15 centimeters), alternating between green and brown, and make sure each is moist. Once metabolic activity has started, you will need to make sure that the pile stays moist for the entire composting process. Since you don't want the pile to be wet (this encourages anaerobic activity), make sure to mix wet material with dry material if necessary. If you are composting in a dry climate, flatten out or make a concave impression in the top of your pile to collect what rain does fall. Similarly, if you are composting where it rains a lot, cover the pile with a tarp or consider making compost in an enclosed bin.

If a pile is too moist, it won't heat properly. You should be able to take a handful from your pile and squeeze just a few drops of water from it, but no more. If your pile does get too wet, then add dry materials or turn the pile. This is hard work, so it is better to get it right in the beginning.

Hot composting will kill weed seeds and pathogens in most cases, but there is no reason to risk adding diseased material or really noxious weed plant material to your pile until you get the hang of the process and can distinguish compost from what we can tell is merely "almost compost." There is a big difference. You have to finish the composting process to ensure that pathogens and weed seeds have been destroyed.

How do you know you have good compost? Test it. You can send compost out to a biological testing lab, but an easier and cheaper home test is to smell the finished product. If it smells bad, like vomit or putrefying matter or vinegar, then it contains anaerobic organisms and their by-products and should not be used. If it smells like ammonia, then it is not finished. In either case, aerate it to change these conditions, and let it sit for a few days before you give it another nose test. You know what fresh soil should smell like; good compost should smell "clean" as well.

You can also plant something in it. Good compost supports plant growth. If there are not enough predators eating the fungi and bacteria, then the nutrients they hold won't be cycled and you will be able to tell by the plant deficiencies.

Compost for the lazy

A modern mix for "instant compost" requires three cubic yards of brown tree leaves and a 50-pound bag of alfalfa meal from an animal feed store. This mix works even better if the leaves are shredded so the bacterial microbes can get right to work decaying it. If you don't have access to alfalfa meal, start with equal volumes of grass clippings and leaves and work from there. If this pile heats up too much, use less grass. If it doesn't heat up enough, use more grass. This assumes moisture and air are adequate. We learned from experience that if you spread fresh grass clippings out and let them dry for a day or two before adding them to the compost pile, they won't mat or smell.

Make your pile in layers starting with 4 inches (10 centimeters) of leaves followed by a layer of the alfalfa meal (or grass) of the same thickness, another layer of leaves, and another layer of meal, and so forth. Water each layer lightly and then add the next. Add sticks and branches as you go along to increase air circulation through and to the middle of the pile.

Once you have accumulated at least the three cubic yards of organic material needed by your army of microbes and other soil food web organisms, they will go to work. Heat will be noticeable in 24 hours. Thereafter you will need to monitor the temperature: it shouldn't get over 150F (65C) or cool down much below 104F (40C). Turning the pile will increase the heat until the pile reaches the mature stage, after which it won't heat up when you turn it. Turning lowers temperatures temporarily until the microbes start working in concert again. Again, water will cool down a pile.

If this sounds like too much work for you, try cool or cold composting: simply pile organic matter in a corner of the yard and leave it. This material will eventually decay, only very slowly; cool composting can take a year or more versus a few weeks or months for hot composting. The end result is compost, however, and as long as it contains the proper set of organisms, it doesn't matter which system you use. Note that worms, beetles, millipedes, and other micro- and macroarthropods will be represented in higher populations in cool compost. It is, therefore, a good idea to keep a cool compost pile going at all times, no matter how energetic you are; the diversity of soil organisms it adds can only help your garden. In the soil food web, higher member diversity means a better ability to eliminate pathogens or control them, either by direct attack or by competition for nutrients and space.

Vermicompost

Processing organic materials through earthworms makes vermicompost, which is almost always bacterially dominated (few if any fungi are involved in worm digestion). Heat is not involved, as this would kill the worms. Instead, the worms (that is, the bacteria inside them) digest the materials and create castings. You can buy special earthworms for this job and buy or make a small bin to keep them in; this can be a simple wooden or plastic box. Just out of the bin, vermicompost has a bacterial dominance; the castings—coated with a polysaccharide as well as carbohydrates and simple proteins—are perfect for supporting good bacterial populations.

Good starting materials for vermicompost include food wastes (no fats or meats), paper, cardboard, leaves, and green grass; or you can use the same materials as you would to start a normal compost pile. If your material contains weeds, thermally compost them first before adding them to the bin; this prevents unwanted seedlings from growing in the worm bin. Any brown materials need to be shredded or otherwise broken up, so the worms can ingest it quicker. With luck, your materials will also include some microarthropods to help physically break down the matter for the worms. Putting your bin outdoors will encourage arthropod and insect activity in it.

Inoculate your soils

It doesn't take much compost to impart life to the soils. To inoculate your soils, put ¼ to 1 inch (0.5 to 2.5 centimeters) of the appropriate compost (fungal, bacterial, or balanced) around your plants. Fungal compost should be applied around trees and shrubs and most perennials; bacterial compost is most appreciated in veggie and flower gardens and lawns (review Soil Food Web Gardening Rules 1 through 4!). Compost can work its magic in the soil in as little as six months. After only that short period of time, new soil life will be evident in the first 6 to 15 inches (15 to 38 centimeters) of the soil inoculated. With this new life comes all the benefits of the soil food web: decompaction, aeration, better water retention and drainage, and increased retention and availability of nutrients. After a year, the soil life will be down as deep as approximately 18 inches (46 centimeters).

Gathering the materials and making a compost pile does take a certain amount of work. The benefits derived from compost, however, are almost incalculable when it comes to managing the soil food webs in your life. Compost is an indispensable soil food web gardening tool.

Chapter 16

Mulch

MULCH IS anything that can be placed on top of the soil to reduce evaporation, prevent weed growth, and insulate plants. Using this definition, plastic sheeting makes great mulch. For our purposes, however, we are only interested in organic mulches, mulches that come from things that were once alive and can be recycled back into nutrients by soil food web organisms. Organic mulches include leaves and leaf mold, aged pine needles, grass clippings, aged bark and wood chips, straw, well-rotted manure (if you must), seaweed, "almost compost," plant remnants, and paper.

New reasons to use mulches

Most gardeners are familiar with the standard reasons to use mulch in the garden. A thick enough layer will smother existing weeds by depriving them of needed sunlight or prevent them from germinating in the first place. Mulches also help give landscaped areas a neater appearance and keep soils cool when there is too much heat; where it gets cold, mulches insulate the soil, and where there are freeze-thaw cycles, mulch is great at preventing premature plant growth by keeping soil frozen. Mulches prevent the soil compaction caused by heavy rains. They greatly reduce evaporation from the soil.

Absent from the usual list of reasons to use mulch is that mulch provides nutrients and a home for certain soil food web organisms, and a good mulch works wonders in imparting soil food web benefits to the soil. For example, worms pull mulch material into underground dens for shredding; the results are nutrient-rich worm castings, more worms, worm tunnels and dens, better water retention, and improved aeration. All manner of micro- and macro-arthropods are able to live in mulches, speeding decay, adding to the soil's organic content, and attracting other members of the soil food web.

We readily acknowledge that mulch is not as effective as compost for adding microbes quickly to the soil food web. Mulch cannot match compost's diversity of soil food web organisms; the decay process has not been completed (and may not even have started), and thus organic mulches lack the variety and numbers of compost's organisms.

We also admit that mulches can result in a feeding frenzy by bacteria and fungi which—if not matched by a feeding frenzy of nematodes and protozoa upon the bacteria and fungi—can result in nutrients being tied up to the detriment of plants in the area. This is another reason mulches control weeds so well: the biology in mulches ties up nitrogen, sulfur, phosphate, and other nutrients on the soil's surface, where the mulch is put down. These are not available to shallow-rooted weeds, while deeper down in the soil, where your plants roots are located, things are fine. When mulches are used properly, however, nutrients can be cycled from them.

The one benefit of using mulches that should be evident to you by now: if you use the right kind of mulch, you can establish dominance of fungi or bacteria.

Bacterial vs. fungal mulch

Rule #6 remains operative here. A mulch of aged, brown organic materials supports fungi; a mulch of fresh, green organic materials supports bacteria. Mulching your garden with brown leaves will encourage a flush of fungi; placing green mulch on soil will foster populations of bacteria. Either will eventually attract microarthropods, arthropods, worms, and other soil food web participants. These will work through the mulch, pulling bits of it into the soil, shredding and tunneling through it, taxiing other members of the web to new locations. You know the routine—a soil food web evolves.

A number of good organic mulches are available free or at low cost. Fresh grass clippings, the most readily available green mulch, contain all the necessary sugars to attract and feed bacteria. Avoid grass taken from lawns where weed killers and pesticides have been applied (and don't accept clippings from yards where dogs are part of the soil food web). Be careful not to pile grass clippings too thick, as they can start to compost and go anaerobic. This will create an offensive odor or heat that can interfere with the very soil food web you are trying to impact.

Our favorite brown mulches are made from the leaves we save each autumn after they fall. These support fungal dominance unless ground up into very fine pieces (in which case they are open to bacteria, who beat fungi into the material). It is also our experience that leaf mulches grow more fungi (or at least grow fungi faster) than do wood chips.

Peat moss is often used as brown mulch. Peat, however, is biologically sterile and should be mixed with other materials to introduce some microbiology.

Pine needles, another brown mulch available to some, make great mulch, but only after they are aged a bit: they contain terpenes, volatile chemicals that are toxic to many plants. Cedar chips also contain high levels of terpenes and should be avoided, but most other wood chips, shredded or chipped bark, and sawdust are great brown mulches and work fine, especially if they are aged or if you mix in some form of organic nitrogen, such as green grass or even alfalfa meal, to ensure the C:N ratio is adequate and nothing need be borrowed by the microbes from the soil under the mulch.

How long mulch will remain effective depends on the kind of mulch used. For example, a 2-inch (5-centimeter) layer of bark chips will last about three or four years, as the lignin, cellulose, and waxes in the bark are difficult for microbes to decay. During this time, fungi will dominate. Leaves, on the other hand, can be completely decomposed in six months; fungi dominate at the start, but bacteria increase once they are able to get inside the material.

Leaves make great brown mulch. Photograph by Judith Hoersting.

Where and how you place mulches also plays an important role. Rule #7 (mulch laid on the surface tends to support fungi, while mulch worked into the soil tends to support bacteria) means it is possible to use one kind of mulch, say tree leaves, and get two different soil dominances. Bury most mulch, and bacteria will have an easier time. If it is on the surface, fungi will dominate the decay activity for a while because it is easier for them to travel from the soil to the mulch.

That is not all. The condition of the mulch is also important. If you wet and grind mulch thoroughly, it speeds up bacterial colonization (Rule #8). Bacteria need moist environments, or they go dormant. And if the material is ground up, it has a lot more surface area; increased surface area means it is easier to get into, and bacterial populations increase. To keep fungi from getting to their food source, some of these bacteria produce antibiotics that suppress fungal growth, making it easier for the bacteria to attain dominance once they get established. If you want more bacteria, use green mulches that have been ground up and soaked. If you only have brown mulch material and need to establish bacterial dominance, chop it into really fine bits and mix some in the top few inches of soil.

On the other side of the coin, coarse, dryer mulches support fungal activity (Rule #9). Mulches with less than 35% moisture are considered "dry mulches." Sure, fungi need some moisture to thrive and grow, but bacteria are more dependent on moisture. If you want fungal activity, use brown leaves or wood chips; don't pulverize them or wet them much; and place them on the surface.

C:N ratio—again

In order to decay, mulch requires air, water, carbon, nitrogen, and the right biology; and once again, the ratio of carbon to nitrogen comes into play. If there is abundant carbon in mulch but not much nitrogen, or a ratio of 30:1 or greater, then the decaying microbes use up the nitrogen in the mulch and, once that is gone, will take nitrogen from the soils touching the mulch.

People make a big deal of this nitrogen "robbing," but it usually occurs only at the thin interface of the soil and the mulch. Although it has a real impact there, it usually doesn't affect the rhizosphere or the bacteria and fungi that reside there. Still, there is no reason to court problems. Experience has taught us that the chances nitrogen will be immobilized in soils under wood chip mulch can be reduced by making sure the chips are $3/8$ inch or larger. This prevents much of the bacterial colonization you would see in smaller wood chips, and—

where mulches are concerned—it is primarily the bacteria that tie up the nitrogen in the surrounding soils.

Applying mulches

Mulches are easy to acquire and relatively easy to handle and use in support of your soil food webs. Simply apply the rules and the appropriate mulch (green or brown; wet or dry; coarse or fine) in the appropriate way (dug in or on the surface) around your plants (vegetables, annuals, and grasses, or trees, shrubs, and perennials). Be careful: add a layer any thicker than 2 to 3 inches (5 to 7.5 centimeters) and you may end up blocking moisture and air and smothering mycorrhizal fungi. Do not put mulch snug up against stems or trunks; this can cause microbial decay of the plant itself, so back off a bit.

If you already use mulches on your property, you know what great things they can accomplish: keeping weeds down, holding in moisture during the summer, insulating soils in winter. They save a lot of work, don't they? Imagine how much more work they will save when you use them to help feed plants the kind of nitrogen they prefer. So correct any mulching mistakes you may have made and reapply the proper kind of mulch, in the proper way, to each plant type you have.

Mulches excel when they are used in conjunction with compost. Put the compost down first and then cover with mulch. As they do the soil, the compost organisms will inoculate the mulch, and begin to decay it as well.

Finally, you can foster all the bacteria and fungi you want in mulch, but if you don't also have the proper nutrient cyclers, specifically protozoa and nematodes, it is not going to have a big effect on your plants. You can actually grow your own protozoa by soaking fresh grass clippings, alfalfa, hay, or straw in dechlorinated water for three or four days. It is a good idea to bubble the water with an aquarium air pump and air stone (available at garage sales everywhere) to keep the mix aerobic. If you look carefully at this soup, you should be able to just make out protozoa dashing around (use a hand lens, and you're guaranteed to). Pour this protozoa soup on mulches, and you will increase the nutrient cycling power of the second soil food web gardening tool.

Chapter 17

Compost Teas

COMPOST TEA—the third tool in the soil food web gardener's shed— puts the microbiology back into soils. This is a good thing because there are some practical problems associated with using the other two tools, compost and mulches. Besides the effort of turning a compost pile, if you have a decent-sized garden and lots of trees and shrubs, carting compost and mulches around and applying them can be hard work. You also have to have lots and lots of both if you are working on anything but a small yard. But the chief problems with these two tools? They take a while to reach the rhizosphere. And neither mulch nor compost sticks to leaves. Plants generate exudates from their leaves, attracting bacteria and fungi to the phyllosphere, the area immediately around leaf surfaces. As in the rhizosphere, these microbes compete with pathogens for space and food and in some cases can protect the leaf surfaces from attack. You cannot immediately introduce this microbiology into the rhizosphere, or into the phyllosphere at all, with compost or mulch.

Actively aerated compost teas, on the other hand, are usually easy to apply—to both soil and leaf surfaces—and are put right where they are needed. They are a fast, inexpensive, and definitely fascinating way to manage soil food web microbiology in your yard and gardens, handily overcoming the limitations of compost and mulch.

What AACT is not

Do not confuse actively aerated compost tea with compost leachates, compost extracts, or manure teas, all of which have been employed by farmers and gardeners for centuries.

Compost leachate is the liquid that oozes out of compost when it is pressed or when water runs through it and leaches out. Sure, these concoctions get a bit of color and may have some nutrient value, but leachates do little to impart microbial life to your soils: the bacteria and fungi in compost are attached to organic matter and soil particles with biological glues; they don't simply wash off.

Compost extract is what you get when you soak compost in water for a couple of weeks or more. The end result is an anaerobic soup with perhaps a bit of aerobic activity on the surface. The loss of aerobic microbial diversity alone (not to mention the risk of its containing anaerobic pathogens and alcohols) suggests that compost extracts are not worth the effort. We don't consider it safe or advisable to use them.

Manure tea, created by suspending a bag of manure in water for several weeks, is also anaerobic. Using manure is asking for pathogenic problems and, especially under anaerobic conditions, virtually assures the presence of *E. coli.* We want the beneficial microbes to be working in our soils and to get these, you have to keep things aerobic.

Modern compost tea

Modern compost teas, on the other hand, are aerobic mixtures. If the tea is properly made, it is a concentrate of beneficial, aerobic microbes. The bacterial population, for example, grows from 1 billion in a teaspoon of compost to 4 billion in a teaspoon of an actively aerated compost tea. These teas are made by adding compost (and some extra nutrients to feed its microbes) to dechlorinated water and aerating the mix for one or two days. It is this mixing, or active aeration, that brings old-fashioned anaerobic compost teas into the modern era; it is also what keeps these compost teas aerobic, and thus safe. The air supply must be sufficient to keep the tea aerobic throughout the entire process.

It takes energy to separate microbes from compost. You know how much energy you have to use daily (or should) to remove another form of bacterial slime: plaque on your teeth. Bacterial slime in soils is just as strong. Consider, as well, that fungal hyphae grow not only on the surface of the compost crumb

10–150 µg	**Active bacteria**
150–300 µg	**Total bacteria**
2–10 µg	**Active fungi**
5–20 µg	**Total fungi**
1,000	**Flagellates**
1,000	**Amoebas**
20–50	**Ciliates**
2–10	**Beneficial nematodes**

Minimum standards for organisms per milliliter of compost tea. Courtesy Tom Hoffman Graphic Design.

Actively aerated compost tea is teeming with bacteria, fungi, protozoa, and nematodes extracted from compost.
Photograph by Judith Hoersting.

but inside its nooks and crannies; you have to use energy to pull these strands off and out in addition to getting the bacteria "unglued." Of course, too much energetic action can kill these microbes. A brewer's action must be strong enough to tease out the microbes but not so strong that the microbes are killed once they are out of the compost and into the tea.

The brewer

More and more compost tea brewers are on the market. These range from small, 5- to 20-gallon systems that can easily make enough tea to take care of a few acres (about 1.2 hectares) to commercial brewers capable of producing up to a thousand gallons or more of tea per brew. The Internet is a good place to look for compost tea brewers and compare them. Manufacturers should be able to show tests demonstrating that their machines can extract viable populations of fungi as well as bacteria. Only a biological test will tell you the numbers. Insist on seeing one, and if they don't have one, don't buy the machine.

You can also make an actively aerated compost tea brewer. It is very easy and our suggestion for those just starting with teas. All you need is one of those ubiquitous five-gallon plastic buckets; add to this an aquarium air pump (the biggest you can afford) and air stone, and about 4 feet (1.2 meters) of plastic tubing to use with it. The better pumps have two air outlets; if you cannot get a double-outlet pump, use at least two single outlet pumps. Sufficient aeration is critical. Once your system is operating, you will know if you have enough air. If the tea smells good, things are fine. If it starts to smell bad, the tea is going anaerobic.

We learned in physics that the smaller the bubbles, the higher the surface to air ratio and thus more air exchange with the water, but when bubbles get

The BobOLator, which uses a chamber to hold the compost, makes 50 gallons of tea in 24 hours. Photograph by Judith Hoersting.

The KIS commercial brewer can make enough tea in 12 hours to treat a one-acre property. Photograph by Judith Hoersting.

too small, under 1 millimeter, they can cut up microbes. Aquarium air stones work well as long as you remember to keep them (and the plastic tubing that attaches them to the pump) clean. Another system can be made replacing the air stone with a two-foot link of $1/4$-inch soaker hose designed for drip irrigation systems. This hose can be coiled and taped onto the bottom of the bucket, giving better bubble "coverage" than an air stone.

Using a bit of duct tape, we tape the air stone or soaker hose to the bottom of the bucket, then connect the tubing and run it out of the bucket to the pump. If you want to have a really good-looking system, you can buy a small rubber grommet designed to be placed inside of the bucket wall so that you can thread the air tubing through it without having liquid leak out. If you put this low enough on the bucket wall, or even in the bottom of the bucket, it is easier to keep whatever you use to create bubbles down on the bottom of the bucket.

Some people put their compost in a porous bag before they put it into the tea brewer rather than allowing it to mix freely in the water. This eliminates the need to strain tea before you apply it, which you will have to do if you are going to use the tea in any garden sprayer (if you are only going to use tea as a soil drench, straining is not a problem). A pair of large-sized pantyhose works well as such a "compost sock." We'll save the male readers some research time: we

It is easy to make a simple actively aerated compost tea brewer using aquarium pumps and air stones.
Photograph by Judith Hoersting.

learned by standing around and reading the labels at the store display that the largest pantyhose are often size Q. You can stretch the waist of a size Q all around the top of a five-gallon bucket, so that the legs fall into the bucket, and drop the compost right in the legs. Or you can tie the legs in a knot and fill the "bag" this creates with compost. It will sit in the water.

Siting and cleaning the brewer

Temperature is important when brewing compost teas. If it is too cold, microbial activity slows. If temperatures get too high, then the microbes are literally cooked or go dormant. Room temperature is ideal. Keep track of the water temperature. This is one of the variables you can adjust later, if need be, and a record of this information will be helpful to the lab testing your samples. If you cannot site your brewer in a warm place with steady temperatures, then a small, inexpensive aquarium heater might be needed; these come with automatic thermostats. If it is too hot where you make tea, you may have to consider "packing" your bucket with ice or occasionally adding ice to it to keep temperatures down.

Compost tea should be made away from direct sunlight because its ultraviolet rays kill microbes. And, since the proteins (worm bodies, primarily) in compost have a tendency to foam in the tea, make sure you keep your brewer in a spot that can tolerate some spillage.

These black rings are bioslime that formed on the inside of a compost tea brewer basket. If allowed to remain, bioslime can detrimentally impact the quality of the tea produced. Photograph by Judith Hoersting.

It should be obvious but must be noted that it is important to clean up right away when making actively aerated compost teas. Bacterial slime is strong stuff and can clog the air holes in bubblers and tubing. This bioslime will appear in the strangest places. It will stick to the sides of the bucket and accumulate in the crevice at the bottom of the bucket. You may have to take apart hoses and fittings to clean them thoroughly. So, even before you use your tea, clean your system. If you get to it while it is still wet, you can usually wipe it off or "blow" it off with the force of water from a hose; at a minimum, flush it with water. Use a 3% hydrogen peroxide product or a solution of 5% baking soda to clean slime that has dried.

Ingredients

Actively aerated compost teas contain lots of bacteria, fungi, nematodes, and protozoa because that's what's in compost. What makes these teas such a good soil food web tool (besides the high concentration of microbes) is that you can tailor-make AACTs to feed plants according to their specific needs by adding certain nutrients (see Rule #10). Use Rule #10, which applies equally to compost, mulches, and soil, when you make compost tea, and it evolves into Rule #11: by choosing the compost you begin with and what nutrients you add to it, you can make teas that are heavily fungal, bacterially dominated, or balanced. For many, the brewing process grows into a hobby in and of itself, not unlike making beer.

All recipes, however, start with the basic ingredients, the first being chlorine-free water. Rule #12 is very important: compost teas are very sensitive to chlorine and preservatives in the brewing water and ingredients. It is vitally important that none of the ingredients you use contain any preservatives. This makes sense. After all, these chemicals are intended to kill or discourage microbial life. If you are served by a water system that uses chlorine, you will need to fill your brewing container with water and run air bubbles through it for an hour or two. The chlorine will evaporate, making the water safe for microbes. Carbon filters and reverse osmosis water systems also work well to remove both chlorine and chloramines, and are particularly useful if you need large quantities of water. As a general rule, a carbon filter containing one cubic foot of carbon will filter four gallons of water a minute.

Next, you need to use good compost (forgive this redundancy: to us, all compost is good, or it isn't compost). Again, make sure there are no chemical remnants in it, and by all means give it the sniff test. If it doesn't smell good, it isn't good compost. Obviously, the best way to know is to have it tested. Avoid "almost compost," compost that hasn't finished the process or has gone stinky and anaerobic. Don't bother with compost that was allowed to overheat, killing beneficial microbes and reducing its soil food web. If you have a low diversity of microbes in your compost, you will have low diversity in your tea.

Vermicastings are a good substitute for compost. These are full of beneficial microbes and tend to be very bacterial (remember the role bacteria play inside the worm, digesting food), especially when they are fresh. For the initial five-gallon brew, you will need approximately four cups of either compost or vermicompost. You can use proportionately less compost the bigger the brew.

As for the extra ingredients, you can feed the microbial population while teas are brewing. Molasses (nonsulfured, so as not to kill the microbes) in powdered or liquid form, cane syrup, maple syrup, and fruit juices all feed bacteria in teas and increase their populations. Two tablespoons of any of these simple sugars in four or five gallons of water will help bacteria multiply and establish

DECHLORINATED WATER	COMPOST
25 gallons	5 lbs (20 cups)
50 gallons	7 lbs (28 cups)
500 gallons	15 lbs (60 cups)

The amount of compost (or vermicompost) used to make tea varies nonlinearly, as this chart shows. Courtesy Tom Hoffman Graphic Design.

dominance. If you make a bigger brew, add more nutrients in the same proportion: the amount of all added nutrients will vary linearly as you increase the size of your brew. More complex sugars and fish emulsion are also good bacterial food, though both will also support some fungal growth.

To encourage fungal growth in compost teas, add kelp, humic and fulvic acids, and phosphate rock dusts, which not only provide the fungi with nutrient value but also give them surfaces to attach to while they grow. *Ascophyllum nodosum* is a cold-water kelp that can be purchased over the Internet, at garden centers, and even animal feed stores, where it is often sold as powdered algae. The pulps of fruits like oranges, blueberries, and apples will also help fungi grow in compost teas, as will aloe vera extract (without preservatives) and fish hydrolysate (which is essentially enzymatically digested ground-up fish—bones and all). You can buy fish hydrolysate at some nurseries or make your own by adding papain (aka papaya peptidase) or kiwi (which also contains the appropriate enzymes) to a blend of fish to enzymatically digest the bones. Yucca and zeolites are also good fungal foods and do not support populations of bacteria.

Give fungi a head start

Many new to tea brewing become frustrated because it can be difficult to grow fungi in quantities sufficient to make a balanced tea, much less a fungally dominated one. This is because bacteria not only grow but multiply rapidly in tea given adequate nutrition; whereas the brew time is almost never long enough for fungi to multiply in tea—they only grow bigger. The better way is to activate fungi in the compost prior to making tea, allowing populations to multiply before they are teased out of the compost and into the tea brew.

This activation is easily accomplished: several days before brewing the tea, mix the compost with simple proteins that serve as a good fungal food—such things as soybean meal, powdered malt, oatmeal, oat bran, or, best of all, powdered baby oatmeal. Thoroughly mix in one of these at the rate of three or four tablespoons per cup of compost. Make sure there is sufficient moisture in the compost, which is to say a drop of moisture can be squeezed out of a fistful of it. Put the mixture in a container, and place the container in a warm, dark place. A seed-germinating mat, placed beneath the container, works great to provide the proper heat.

After about three days at 80F (27C), the fungi in your compost, if you had sufficient numbers of them in the first instance, will have grown, and their invisible hyphal threads merged into a network of visible mycelia. The compost

Fungal mycelia are activated by adding fungal nutrients to compost before making tea.
Photograph by Judith Hoersting.

should look like Santa Claus's beard, covered with long, white, fluffy strands. In a few more days, there will be so many fungal threads, the entire container of compost will be glued together.

Teatime

Once you turn your machine on, the bubbles agitate the compost and start peeling microbes off and out of it. Depending on the compost and the nutrients, you may experience a bit of foaming; this can signal that worm protein is being released from the compost—a good thing. You can add mycorrhizal fungi at the very end of the brew cycle. If you put spores into the tea while it is being made, either they will be destroyed or the fungal hyphae they produce will be destroyed—they are both very fragile; also, since mycorrhizal fungi live off of root exudates, they and the tea must reach plant roots quickly.

It takes between 24 and 36 hours to develop a good tea using our simple bucket bubbler; some commercial brewers, with their high-energy systems, make tea in 12 hours. In any case, during the course of the brewing, tea turns coffee-brown, another favorable sign: the humates in the compost are being teased out into the tea. The temperature of the brew may also increase a few degrees, a result of increased metabolic activity. The best part is the smell. The smell of compost teas, especially when molasses is used as a nutrient, is a healthy, sweet, earthy smell.

Compost tea has a very short shelf life. So many microbes now populate the brew that they quickly deplete the nutrients and start eating each other; more important, they are using up all the oxygen. If you are offended by the odor of a tea, it has probably gone anaerobic and should be discarded; do not toss it on your plants, for obvious reasons. It is best to use compost tea within four hours of manufacture, though it will last, diminishing in populations, for about three to five days if kept refrigerated or if you continue to bubble air through it.

After you have had some experience making teas, you may want to modify your machine in order to make better and better teas, meaning those that have higher numbers of microbes. For example, besides substituting the soaker hose for the air stone, we also upped the size of our pump; eventually we found a used, $1/3$ horsepower air pump, and now make seriously bubbling tea in a 30-gallon plastic garbage can (affectionately known as the "Lawrence Welk–o-Lator"). The bubbles come from various pieces of equipment; we are continually experimenting, using specialized fish tank and Jacuzzi aerators, watering can heads, and even a plastic water pipe pocked with holes made with $1/16$- and $1/8$-inch drill bits.

Application

Right at the outset we will tell you that you can never apply too much compost tea (our research shows no ill effects from unlimited applications). It doesn't burn plant roots or leaves, and the microbiology in the tea will adjust to the nutrients available at the site. Repeatedly applying compost tea will only help increase diversity of the microbial populations in your soils. Use tea on lawns, vegetables, trees, shrubs, annuals, and perennials. Unlike chemical sprays and soaks, compost tea is safe and easy to apply.

Once the tea is ready, apply it as a soil drench using a cup, a plastic watering can (bacteria can impact the zinc in metal containers), or (if the tea has been strained) a hand pump sprayer. Since compost teas will "stick" to leaf surfaces, you can inoculate leaves with a foliar spray of beneficial microbes. To be effective as a foliar spray, the tea must cover 70% of the leaf surface. Cover both sides of the leaves. When applying compost teas to soils, drench your plants and the area around them with the tea. You cannot overdo it.

And don't forget the sun: ultraviolet rays kill microbes. If you live in southern latitudes, you will want to apply before 10 a.m. or after 3 p.m., when UV rays are weakest, even on a cloudy day. There is no microbial sunblock lotion. It can take 15 to 30 minutes for bacteria or fungal hyphae to attach themselves to a leaf (where they can get some protection)—far too long a period to be exposed to the sun's rays. Alternatively, spray with a drop diameter of at least 1 millimeter; with that much water, bacteria can develop enough slime to establish themselves before the water even evaporates. UV rays can also negatively affect the microbiology in soil drenches, but you can be a bit more relaxed about the timing of these since the microbes sink into the soil and leaf duff layer almost immediately.

Remember, you are dealing with living organisms here. The microbes you carefully cultivated and nurtured in your tea are very much alive and require gentle treatment. Sprayers must not exceed pressures of 70 pounds, and the velocity of the spray should be slow. Either stand back or turn the spray head up, so that the tea drops "parachute" down to the surfaces to be covered; there should be no forceful "splatting" of the tea onto the soil or lawn or plants, as this is what will sometimes kill the plant, not the pressure of the tank. Electrostatic sprayers, incidentally, may destroy microbes by putting the wrong charge on them, so test the tea from such a sprayer before using one.

It is possible to use a hand pump sprayer if you strain your tea, but you must take care not to strain the microbes out. The mesh of any "compost sock" should be at least 400 micrometers, which is big enough to let fungi and nematodes flow through but will keep out particulate matter that will clog conventional sprayers. Alternatively, you can decant a tea solution by letting it sit for 15 minutes after the aeration is stopped. This gets rid of a lot of the bits and pieces; the bad news is that often the amount of fungi in the tea is diminished.

You will be better off if you invest in a concrete sprayer, which is capable of handling the particles of compost that would clog a normal garden sprayer. Concrete sprayers look exactly like home garden pump sprayers, only with fewer bends, larger orifices, and nozzles that support bigger particles. For prices and availability, check with your local builders supply store, concrete contractor, concrete supply store, or sand and gravel company. A gasoline backpack mist sprayer is also appropriate, especially for a large yard. A great way to do a lawn is to use a traveling sprinkler with a fertilizer dispenser feeding tea into the water stream (see chapter 18 for more details).

Whether sprayed or poured, the microbes in the tea will establish themselves, grow, breed, attract predators, eat and be eaten, or go dormant. They create protective barriers around the roots and release nutrients when they die. They create and improve soil structure. They make protective barriers on leaves and compete with bad guys there as well.

Compost teas go to work immediately, and for this reason it is important that the tea applied be a good one, full of beneficial organisms, not diseases or pathogens. There is little room or tolerance for a poorly made tea. If you are not up to the job yourself, you can purchase AACTs from an ever-growing number of commercial nurseries and garden centers; some companies not only make but will apply compost teas for you. In either case, it is still advisable to ask for tests to see how the tea measures up and, of course, don't be afraid to give commercially made teas the smell test before buying or applying them. They may have started out fine but gone anaerobic before sale.

You can apply AACTs as often as you like, but how often you *need* to apply them (especially if you are paying for them) depends, as you can imagine, on the status of the soil food web organisms in the areas concerned. First-timers should get a base reading on microbiology and arthropod counts before "taking up" this very effective tool. As your soil food web becomes healthier, you'll need to apply tea less often. Thus, if your yard has had applications of chemical fertilizers for years, you should put down compost tea every other week for three months to establish a healthy soil food web population. Then you can start applying tea once every month for a season and, finally, three times a year.

How much compost tea should you apply in any given session? For two years one of us used about 60 gallons a week on a quarter-acre lot with positive results (save for a few complaints from a spouse that felt too much time was being spent teaming with microbes). The general rule, however, is to apply five gallons of compost tea per acre as a soil drench, ten gallons if you are going to spray leaves as well. It is fine to dilute the tea; just make sure there were five gallons when you started. When you are more experienced, you can match the amount of tea you apply with soil tests and tea tests to achieve specific fungal or bacterial ratios.

Timing

There are certain times when it makes even more sense to apply a tea. For example, it is a good idea to apply teas immediately after leaves fall in the autumn. If the soil and leaf litter don't freeze in the winter, decay will proceed apace all winter long. Even with snow cover, decay will occur at the interface of the snow and the soil surface, where it will warm up enough for microbial activity to continue. Come spring, just before plants start their new growth, put down tea again: ten gallons of soil drench per acre is our suggestion. Treat opening buds and young leaves to a foliar spray of five gallons per acre, as well. If your plants are thriving and are disease-free, you need apply tea only at these

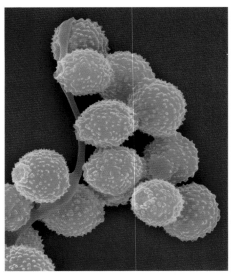

Powdery mildew growing on leaves.
Compost tea sprayed on leaves can out-
compete this and other fungal diseases.
Courtesy Clemson University, USDA Cooperative
Extension Slide Series, www.forestryimages.org.

Powdery mildew up close. Image copyright
Dennis Kunkel Microscopy, Inc.

two times; if you live in a tropical environment, you should apply tea four times a year.

When it comes to outcompeting disease organisms in the soil or phyllo-sphere, fungally dominated teas have been used to prevent and suppress the growth of powdery mildew (*Erysiphe graminis* on turf, *Phytophthora* spp. on rhododendrons), downy mildew (*Sclerophthora* spp.), take-all (*Gaeumanno-myces* spp.), gray snow mold (*Typhula* spp.), pink snow mold (*Microdochium* spp.), red thread (*Laetisaria* spp.), crown and root rots and damping off (*Pythium* spp.), brown patch (*Rhizoctonia solani*), summer patch (*Magna-porthe* spp.), rusts (*Puccinia* spp.), and fairy rings (all sorts of fungi).

Bacterially dominated teas have been useful in outcompeting pathogens in mild cases of dollar spot (*Sclerotinia* spp.—severe infestations also require lots of fungal competitors), necrotic ring spot (*Leptosphaeria* spp.), yellow patch (*Rhizoctonia cerealis*), leaf spots (*Bipolaris* spp., *Curvularia* spp.), pink patch (*Limonomyces* spp.), and stripe smut (*Ustilago* spp.). Insects too succumb to the effects of compost teas, specifically weevils, grubs (*Ataenius* spp.), cut-worms, and chafers; several reports attest to negative impacts on whiteflies, fire ants, and scale.

At the first sign of disease or insect infestations on any of your plants, ap-ply teas and repeat in five to seven days. Obviously, a prophylactic application

Root rot and damping off (shown here on bent grass host) can also be controlled by applications of actively aerated compost tea. Courtesy Clemson University, USDA Cooperative Extension Slide Series, www.forestryimages.org.

is best: if you have a sense of your yard's phenology (seasonal cycles), you should be able to apply teas in advance of breakouts.

Finally, certain weeds are affected by compost teas. Clover and quack grass have a tougher time of it when you add lots of protozoa and beneficial nematodes to the soil; this teas do and increase nitrogen cycling. Plantains, chickweeds, and nut sedges disappear if you reduce the nitrates in soils: use a fungally dominated tea. Ivy also responds to highly fungal teas.

Compost teas are a veritable liquid soil food web. Instead of lugging around wheelbarrows of compost, consider compost teas, a concentration of the same microbiology. When you use them, you are really teaming with microbes.

Chapter 18

The Lawn

USED TO BE, if you were not happy with the way your lawn looked, you put down manure or top-dressed with compost. If you had weeds, you or your children eradicated them by hand. All that changed in 1928 when a company that sold grass seed came up with a way to make synthetic, nitrogen-based fertilizers and to do so cheaply. The rest is history: through aggressive advertising, and let's face it, fantastic results, the chemical side of lawn care has grown into a multibillion-dollar industry.

A vicious cycle

Chemical lawn fertilizers work, and they work well. Their concentrations of nitrates are so high, they are immediately effective: fertilizers are chemicals that feed the roots directly, bypassing the biology in the soils. However, applications of synthetic fertilizers kill off most or all of the soil food web microbes (Rule #13). These fertilizers are salts, and when they come into contact with soil microbes, they cause osmotic shock—that is, water in the cells of these organisms flows to the higher concentration of salts without, literally bursting through cell walls and killing off the microbes that hold (bacteria and fungi) and cycle (nematodes and protozoa) nutrients.

How quickly a lawn's soil food web organisms are affected by chemical fertilizers depends on the organisms in question, their concentration and strength, and the amount of fertilizer applied. A good rule of thumb, however, is that 100 pounds of nitrogen lawn fertilizer per acre will wipe out a healthy soil food web. Lesser quantities kill fewer members of the soil food web, but do damage it nonetheless. What isn't killed outright by four 25-pound bags of lawn fertilizer is driven from the acre by its lack of food resources or by the odor of the chemical fertilizers themselves. When microbiology is missing, as you know, you have to apply (and reapply) the nutrients necessary to keep the grass green.

With the natural buffering action of bacteria and fungi lost, a soil's pH is thrown out of whack; soil pH gets lower and lower as more nitrate salts are applied, eventually requiring readjustment. Matters are made even worse by the common practice of removing grass clippings while or immediately after mowing. The chemical gardener is usually one that "cleans up" after mowing, and even the organic gardener all too often has the knee-jerk urge to rake grass clippings. By removing clippings and autumn leaves, a gardener unwittingly compounds the destruction of life in the lawn's soil. Then again, if you don't have a soil food web to break down and decay leaves and clippings, you are compelled to remove them so they won't block the light the lawn needs.

The use of chemical fertilizers sets off a vicious cycle, then: the more fertilizer you use, the more the soil food web is destroyed, and the more fertilizer you'll need to fill the nutrient void you've created. It is a downward spiral. The end result is either a lawn in really terrible condition or a gardener who has to do a lot of work. Removing the clippings from and applying salts to a lawn leaves the gardener, alone, to do all the work that was formerly carried out by the trillions upon trillions of microbes who used to be on the job. Earthworms leave the area when salts are applied; salts are irritants, and the gut microbes responsible for worm digestion die if fertilizers are ingested. The fungi that bind soil aggregates are gone. The bacteria that produce the slime that binds individual soil particles into aggregates are gone. The lawn's soils lose structure. Slowly, they lose the ability to hold air and water. It is soon Katy-bar-the-door time, and more diseases and problems will arise.

Without a well-populated soil food web, natural defenses are gone. Lawns infested annually with mildew, black spot, rots, gray mold, and other disease-

Dollar spot, one of the two most troublesome diseases of golf course greens, can be caused by excessive nitrates in chemical fertilizers. Photograph by Kevin Mathias, USDA-ARS.

causing opportunistic microbes clearly lack the diversity of beneficial organisms that would normally keep these things in check. By teaming with microbes, you can have a healthy and attractive lawn—with a lot less work on your part.

Taking stock

As with any other area of the yard, it is important to first determine the status of your lawn's soil food web. Biological soil tests by a competent lab are the only accurate way to learn what needs to be corrected and exactly how much restoration work you have to do, but other things will give you a pretty good indication of its state. Earthworms, for instance, won't be present if there are no bacteria, fungi, and protozoa to eat; their presence, therefore, is an excellent indicator of a healthy food web. If you have a good population of worms, your lawn already has lots of beneficial organisms building soil structure, cycling nutrients to the grass roots, building water- and air-retention and drainage capacity, and fighting pathogens. So, if you see birds hunting for earthworms, lots of earthworms after a good rain, or worm castings deposited on the lawn's surface at night, you probably have only to maintain the lawn's soil food web, not add microbiology to establish one.

A lawn maintained by the soil food web. Note the yellowish back area, which was not treated. Courtesy Soil Foodweb, Inc., www.soilfoodweb.com.

Similarly, your lawn's soils should contain plenty of microarthropods—the little arthropods you need a hand lens, MacroScope, or light microscope to see. These help with nutrient cycling, open up the grass clippings, and help aerate the soil. Use a Berlese funnel; if you discover that your soils are lacking these members, you can restore the microbiology by providing beneficial fungi, bacteria, protozoa, and nematodes—the base that will attract arthropods, worms, or other soil food web participants that are missing.

The care and feeding of microbes

At the beginning or end of the growing season, spread an organic fertilizer (microbe food, really) on your lawns. This will ensure that there is a sufficient supply of organic matter to feed the microbes in the soil. Microbe food? This is a big but necessary change in gardening terminology. When you team with microbes, you feed them, and they feed the roots.

Rule #14 warns that if you want to work with the soil food web, you need to stay away from additives that have high NPK numbers. Most gardeners know these letters represent the percentages of nitrogen, phosphorus, and potassium in the fertilizer, and this NPK trilogy appears on all fertilizer packaging. Don't put anything on the lawn with NPK numbers greater than 10-10-10; traditional organic fertilizers usually meet this criterium. Of particular note is that a high (anything over 10) concentration of phosphorus not only prevents mycorrhizal fungi from growing but kills off the ones that are there. As a result, the grass loses its ability to take up a resource easily, and no matter how much

Mycorrhizal fungi (see bowl on the right!) help lawns grow. Courtesy Mycorrhizal Applications, www.mycorrhizae.com.

phosphorus you put on the lawn, it is locked up quickly and unavailable to the mycorrhizae-less grass plants.

Our favorite microbe food for lawns is soybean meal with an NPK of 6-1-1. This is applied at a rate of 3 or 4 lbs per 100 square feet. Other useful organic microbe foods include alfalfa meal, blood meal, cottonseed meal, feather meal (all applied at the rate of 4 lbs per 100 square feet at first and then adjusted to taste) and fish bone meal (3 lbs per 100 square feet—but we warn you, there will be a heavy fishy smell for a few days). These all feed the soil biology; they are not absorbed by plant roots—hence, microbe food, not fertilizer.

It also helps to encourage a suitable environment for the lawn's microbes. We know from Rule #2 that lawns prefer slightly bacterially dominated soils. For this reason alone it is a good idea to leave grass clippings on the lawn, all season, as a bacteria-favoring mulch. The sugars in the grass will attract a healthy population of bacteria. Clippings also foster populations of protozoa, which ensure nutrient cycling. And you will have to mow less, now that high amounts of concentrated nitrates are not being sucked up into by plant roots.

When leaves drop at the end of the season or when twigs and small branches fall after a storm, do not rake them. Instead, mulch them up in place by running your lawn mower over them once or twice. This will open them up and make them available to the fungal components of a lawn, which are also important; fungi help provide structure and drainage and help with the harder-to-digest grass stems that can build up to a thatch layer in their absence. This is why you should rejoice when you see mushrooms in your lawn. They are usually a sign that things are healthy beneath the green grass.

Lawns that have not had the benefit of a healthy soil food web (which may be attributable as much to poor drainage as to chemical fertilizers and weed killers) should be plug-aerated, a procedure wherein 2-inch-long plugs of soil

A handful of plugs pulled from a lawn during aeration.
Photograph by Judith Hoersting.

are pulled from the lawn, creating holes throughout. These holes open up the lawn, allowing water, air, and organic food to enter the root zone. The plugs should be left on the lawn and allowed to decay.

Plug aeration in the early spring every three or four years will help the soil food web because it helps repair compaction caused by the weight of snow and ice or the back-and-forths of pets, children, and vehicles. The aeration is particularly useful in keeping the lawn's fungal population healthy: as the most fragile, fungi are also the first soil organisms to go when a lawn becomes compacted, as it inevitably does. After this spring aeration, apply an organic microbe food. This will fall into the plugholes and provide food down in the lawn's root zone.

Next, inoculate the lawn with beneficial microbes to put microbiology back into the soil or to maintain what is already there. If the lawn is small, this is easily accomplished by applying a thin (up to a half-inch) layer of bacterially dominated compost to the lawn with a fertilizer spreader. If the lawn is large, apply a slightly bacterially dominated compost tea (see "Applying Compost Tea to Lawns" later in this chapter).

What about chlorine in the water you use to water your lawn? It shouldn't affect microbes if you water using a sprinkler. The fine mist spray and the trip from the air to the ground helps clear most of the chlorine from this water. Of course, you can buy an inexpensive chlorine filter and install it on the outside hose bib. One filter should last all season, but you should check the output occasionally to be sure.

Weeding the soil food web way

Lawn weeds can be influenced by the soil food web. Dandelions, for example, appear in calcium-poor soil surfaces. Their long taproots seek out the calcium they lack, and the calcium is deposited in the soil when the dandelion dies. In time—unfortunately, sometimes quite a long time—the soil food web biology works this calcium into the upper layer of soil, where it has been missing. In essence, dandelions can mine themselves out of existence. To get rid of dandelions sooner, boost fungal activity in the soils; fungi tie up calcium, much more so than do bacteria. You can also use a microbe food, corn gluten (a by-product of corn starch production), as an organic, preemergent agent. Put it on lawns with dandelions or other weeds just as they are coming to seed, and it will prevent the new seeds from developing secondary roots. In the meantime, its 10-10-10 formula feeds the soil food web.

Lots of clover or quack grasses in a lawn indicates that the soil food web is not cycling enough nitrogen. Adding nematodes and protozoa via compost, compost tea, or a protozoa soup can increase nitrogen cycling. Chickweed, a frequent weed in lawns, thrives when there is too much nitrate, which is what you get when you put down a commercial lawn fertilizer. Stop applying chemical fertilizers; instead, use the soil food web tools to increase the fungal biomass (and hence the available ammonium) in your lawn.

Moss, on the other hand, indicates that your lawn soil is already fungally dominated instead of being slightly bacterial, as lawn grasses prefer. Mosses like acidic conditions. Apply very bacterial teas and a thin topdressing of very bacterially dominated compost to moss-infested lawns, and the pH will gradually change to one "acceptable" to grass and not as "acceptable" to moss. This will lessen and eventually prevent the appearance of new moss. You should remove the existing moss with a thatching rake and may have to apply iron to kill it first.

As a "soil food webbie," you already know you should be happy to see mushrooms in your lawn. Not too many, of course, which would mean you need to apply a bit more bacterial tea. If you are worried about fairy rings, for example, just increase the diversity of the fungi in your lawn soils by making sure your teas and compost have a good diversity of fungi; the fairy ring fungi should then be outcompeted. In addition, recognize that micro- and macroarthropods as well as mice and shrews eat these and many other fungi, keeping them in check.

Fairy rings and other monocultures of fungi in the lawn can be overcome by increasing diversity with compost or compost tea. Courtesy Clemson University, USDA Cooperative Extension Slide Series, www.forestryimages.org.

Easy changes and good starts

You can use the soil food web to your advantage when it comes to changing pH. Normally you would have to put down hundreds of pounds of lime, gypsum, or sulfur to alter soil pH a few points in a decent-sized lawn; lime in particular acts slowly, taking a season to effect even a point's change. However, you can use considerably less (about one-quarter the amount) and take less time to get the same results by applying some soil food web science. Instead of putting it directly on the lawn, mix lime in when you are making compost. It will be tied up by the microbes in the compost and released during the normal food web cycling. You can put this compost directly on the lawn or you can make compost tea.

Obviously, if you are just putting in a lawn, you have an opportunity to establish a healthy soil food web from the very start, sparing your lawn the indignity of chemical addiction. Before you broadcast grass seed, mix it with the type of endomycorrhizal fungal spores associated with grass plants, vesicular-arbuscular mycorrhizae (VAM). A healthy lawn should have a good portion of roots colonized by VAM for the lawn as a whole to get the benefits of the mycorrhizal relationship. VAM colonization helps grasses compete with weeds for nutrients and blocks root-eating nematodes. And mycorrhizal fungi bring both water and nutrients back to the roots. Biological testing labs can tell you how much VAM you have in your existing lawn soils.

Twenty-four hours prior to seeding a lawn, roll wet grass seed in VAM and store it in a dark, cool spot. VAM will help achieve a healthy lawn that does not need watering or feeding as frequently as those without mycorrhizal fungi.

What if you need a quick fix?

Some lawns are seemingly hopeless, and while soil food web management eventually prevails, quicker action is sometimes desired. Consider first the use of heat, vinegar, or manual labor to get rid of weeds in lawn; if weeds are so bad that you need to use a herbicide, or if the lawn needs an instant nitrate greening (say, for an emergency backyard wedding), then you should take remedial action to restore the soil food web.

Always practice Rule #15: follow any chemical spraying or soil drenching with an application of compost tea. Give the stuff a few days to work, and then apply the tea. The microbes in the tea will immediately start to detoxify the soil by breaking down the remaining chemicals and repopulating it. Repeat in a week, and check the status of soil food web life.

Both bacteria and fungi can degrade pesticides, but it is mostly the fungi that attack and break up these complicated chlorinated carbon rings. You therefore need to inoculate contaminated soil with lots of organic food resources with complex proteins (the kind fungi like), such as kelp, fish hydrolysate, and humic acids.

Applying compost teas to lawns

One of the best ways to establish the right biology in lawns is to use a slightly bacterial aerated compost tea at a rate of five gallons per acre. We are the first to admit that applying compost tea to a large lawn can be problematic if you don't have the right equipment. A commercial tea sprayer service is the easiest way but can be more difficult to arrange and much more expensive than applying it yourself.

Concrete sprayers (see chapter 17) are fine for a small area. For larger areas, you should consider a traveling sprinkler (one that follows along a hose laid out on the lawn) with an inline fertilizer dispenser (a tank made for applying soluble fertilizers) attached to your water source. Instead of holding fertilizer, the dispenser can be filled with actively aerated compost tea, which it will feed to the sprinkler as it travels across the lawn.

A commercial tea sprayer service pays a call. Photograph by Judith Hoersting.

A traveling lawn sprinkler and a fertilizer dispenser make applying tea to a lawn easy work. Photograph by Judith Hoersting.

If you plan on applying tea to a really large lawn, you might want to consider renting or buying a gas blower (and using its lowest, most gentle setting). You can mist an acre of lawn in about five to ten minutes and spray up into 30-foot trees. Rental is the best idea, as you will only need applications in the spring and autumn once the soil food web is established. Do ensure the tank is free of any residual herbicides, pesticides, or other harmful chemicals.

Once your lawn has a thriving soil food web system, it will be much easier to care for. You will no longer have to thatch or rake clippings or leaves. You will need to water less, mow less frequently, and best of all, have the satisfaction of being able to play and work on your lawn without worrying about dangerous chemicals.

Chapter 19

Maintaining Trees, Shrubs, and Perennials

Trees, shrubs, and perennials are the mainstays of any yard's landscaping. Yet they seldom get specialized care and are instead lumped in with the lawn. Whatever fertilizer goes on the grass is usually all the trees and shrubs receive, and all many perennials get as well. The roots of trees and shrubs and some perennials run under the lawn, and they are affected by traffic and by the use of nonselective herbicides, which besides killing lawn weeds kill even the beneficial organisms that protect plants. With a diminished soil food web, you have to become their defenders and continue to feed trees, shrubs, and perennials.

Trees, shrubs, and perennials prefer fungally dominated soils

Ever wonder why the lilacs never bloom? or why that spruce didn't survive when you planted it in the middle of your beautiful green, nitrate-fertilized lawn? Remember, Rule #3 dictates that trees, shrubs, and perennials prefer their nitrogen in the form of ammonium, not nitrate. This means fungal soils. Lawns, on the other hand, do best with nitrates or slight bacterial dominance and therein lies the problem. If the soil is very heavily bacterial, many trees have a difficult time establishing themselves.

Being surrounded by lawns may not be a good thing for trees, shrubs, and perennials—or the gardener—unless some soil food web management practices provide a different soil food web specifically where they grow. We realize that trees and shrubs in particular often function as specimens in the landscape, and that a conifer, for example, that craves ammonium nitrogen might be sited in the middle of a lawn that prefers nitrates. The trick, then, is to try to create an island around each tree and shrub with a fungally dominated soil food web.

The few exceptions to Rule #3 are the trees and shrubs normally considered transitional in the successional development of ecosystems from desert through old growth forest. The most familiar of these are cottonwoods, birches, and

Trees growing in a bacterially dominated or balanced environment should benefit from mulch that will attract fungi. Photograph by Judith Hoersting.

aspen. These do well in bacterially dominated soils when they are young because at that stage of their development they can easily utilize nitrates. Once mature, however, even these prefer ammonium nitrogen.

Trees, shrubs, and perennials dislike compacted soils

Trees, shrubs, and perennials are frequently the victims of compacted soils, especially when they are planted in lawns (as is often the case with trees and shrubs) or in pathed gardens (as with perennials). Every precaution should be taken to prevent this condition (and every step taken to correct it), as roots (and thus plants, obviously) do best in soil with good structure, and good soil structure as you now know absolutely requires an active soil food web.

Larger organisms cannot survive in compacted soil—they cannot move through it in search of food because transportation pathways have been destroyed; if the compaction is really severe, it may be impossible to establish new ones, or not worth the bother. With the nematodes and many of the protozoa gone, nutrients accumulate in fungal and bacterial biomass instead of being released and available to plants. At the same time, the fragile mycorrhizal fungi associated with the roots of trees, shrubs, and perennials are literally crushed or drowned; mycorrhizal fungi that compete with *Pythium* and *Rhizoctonia*, two fungi that cause stem and root rot problems, for example, are missing. After a while, the only soil food web organisms left are the bacteria and opportunistic fungi and protozoa that are so small in size they are able to move through even compacted soil. The food web is not in good shape and surely not full of fungi as trees and shrubs prefer.

Plant roots too have trouble moving through compacted soil. And since they can no longer rely on mycorrhizal fungi to bring back nutrients, plants face a double whammy in compacted soils: they not only don't get the kind of nitrogen they prefer, but their access to water and phosphorus and other nutrients is limited. They become even more stressed.

It gets worse. Compaction reduces oxygen levels, and anaerobic bacteria take over. Anaerobic bacteria produce metabolic products that kill roots. The tunnels and burrows through which water flows, pulling and pushing air, disappear. No mycorrhizae, no beneficial fungi, harmful elements galore—this is not a healthy situation.

Plug aeration of the affected area is only a first step toward remediation of compacted soils. If you don't have the proper soil food web organisms to improve compacted soils, the benefit of aeration will be short-lived. The solution is to apply food web management practices and return the organisms that are needed to build and maintain soil structure. Mulches, compost, and compost teas are all very effective when it comes to treating compacted soil around trees, shrubs, and perennials.

All three soil food web tools apply

Brown mulches and fungal compost and compost tea work best when caring for trees, shrubs, and perennials. Start with compost and place it under all trees and shrubs and around all perennials to a depth of 1 to 2 inches (2.5 to 5 centimeters). Go at least out to the drip line of the tree or shrub, but make sure the compost doesn't touch the stem or trunk of any of these plants (so that, again, the microbes in the compost don't attack the bark). Obviously, you should give up on trying to grow grass under trees.

Gravity isn't the only reason trees and shrubs drop their leaves where they do. The nitrogen and carbon in these leaves is naturally recycled, and some makes its way back into the plant. Nature places mulch over tree roots; you should too—again, to at least the drip line—using brown mulches. Mulch even if you don't have compost to put under your plants. Start with the plant's own

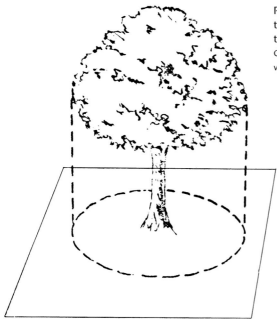

Place compost and mulch under trees and shrubs at least as far as the drip line. Diagram by Tom Hall, Georgia Forestry Commission, www.forestryimages.org.

leaves if you can (open these up for bacteria and fungi by running them over with a lawn mower); don't remove them. Add to nature's mulch with brown mulches of any kind, but don't let it get too deep. A few inches is all that is needed to support a healthy population of fungi. The mulch has the added benefit of keeping down weeds and grass by blocking the light.

Finally, consider an application of compost tea around trees, shrubs, and perennials, once at the beginning of the growing season (two weeks before trees and shrubs leaf out) and again at the end, just as leaves finish falling and are in place under the plants. The microbes in the tea will really speed up decay during the winter months and support a good, fungally dominated food web community. You can simply soil drench, and don't need to bother with sprays, except for perennials, which in addition to the two soil drenches should be sprayed at least once after their leaves appear to add microbiology to the phyllosphere.

Mycorrhizal relationships

Before planting trees, shrubs, and perennials, inoculate them with mycorrhizal fungi. These can be purchased at nursery centers. Remember, there are two basic types of mycorrhizae—those associations where roots are invaded, and those where they are not—so it is important you get the right ones. Which mycorrhizal fungi to use on what is answered by Rules #16 and #17: most conifers and hardwood trees (birch, oak, beech, hickory) form mycorrhizae with ectomycorrhizal fungi; most shrubs, softwood trees, and perennials form mycorrhizae with endomycorrhizal fungi. These rules are based upon the research of soil scientists, who now have the tools to assess what types of fungi naturally associate with particular plants and have codified these assessments. There are exceptions to these rules. For example, plants in the heath family, which includes rhododendrons, azaleas, and blueberries, require ericaceous mycorrhizae, which are not yet commercially available. Nonetheless, if you stick with these rules, you should be on stable (but, we hope, not compacted) ground.

Mycorrhizal fungi spores must come into direct contact with roots within 24 hours of being exposed to moisture in order to grow. Commercial preparations that contain mycorrhizal fungi are always dry powders or grains (mixed with various materials to help in their delivery), so they are easily applied when

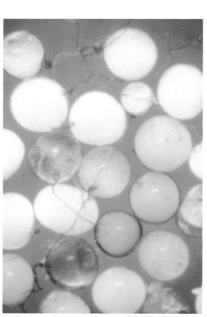

Mycorrhizal spores. Courtesy Mycorrhizal Applications, www.mycorrhizae.com.

The pine on the left was treated with mycorrhizal fungal spores when planted; note the increased size of both the seedling and its root ball. Courtesy Mycorrhizal Applications, www.mycorrhizae.com.

plants are about to be put into the ground. Simply sprinkle them on the roots or dip the roots directly into the spores before you plant, and then water in the new plant as usual.

Existing trees and shrubs are a bit more difficult to colonize. Let's hope your soils have not been degraded to the point that natural mycorrhizal fungi have been affected. Look for signs of mycorrhizae in the form of one particular kind of mushroom growing near the same kind of tree. Birch trees, for example, often form an association with the fly agaric, *Amanita muscaria*. If your existing trees have mushrooms under their drip line, you most probably are looking at an existing mycorrhizal association and don't have to add to create one.

If you have a yard with really compacted soils, have not seen mushrooms around your trees and shrubs, or have noticed they are not doing well, consider using a root feeder or long syringe (the kind used to apply glue) to inoculate the roots of existing plants with the appropriate mycorrhizal fungi. In the case of most perennials and shrubs, you can carefully dig into the root zone with a spade or trowel and apply endomycorrhizal spores whenever you come across roots.

Birch trees often form mycorrhizae with the mushroom *Amanita muscaria*. Photograph by Judith Hoersting.

Unstressed plants are healthier

Stressed trees put out a signal recognized by aphids and other insects; they know the tree is weak and attack it. Unstressed trees don't emit this message, and they are able to produce extra pitch and sap to trap any invading beetles. Their exudates attract all the right microorganisms. Their leaves are coated with beneficial bacteria and fungi to outcompete disease. Their roots have formed mycorrhizae that increase their reach and allow them to dine on phosphorus and wash it down with ample water.

The bottom line when it comes to caring for your trees, shrubs, and perennials: try to plant them in soils that are already fungally dominated. If not, apply fungally dominated compost, mulches, and teas to and around them. Let leaves remain under the plants from which they drop. And, obviously, use all three soil food web tools, especially compost teas, at the first sign of any diseases.

Chapter 20
Growing Annuals and Vegetables

A WHOLE INDUSTRY is built around fertilizing annuals and vegetables. The lawn might be the number one dump for chemical fertilizers, but homegrown tomatoes and marigolds are not that far behind. The same high concentrations of soluble nitrates that work on the lawn, with their percentages tweaked a bit, work quite well when it comes to feeding flowers and vegetables; and the vicious cycle that develops in lawns treated with chemical fertilizers will also occur in your flower and vegetable beds. The natural cycling of nutrients ends. You have to feed the plants you grow with increasing amounts of chemical fertilizers because no longer are there microbes to provide them with nutrients, and in the absence of microbes, soil structure deteriorates. Without a healthy soil food web, opportunistic pathogens and animals appear, and these seemingly require other chemicals to keep them at bay or in balance.

Annuals and vegetables prefer bacterially dominated soils

What are the soils in your vegetable and flower beds like? Look for earthworms. They survive by eating protozoa and bacteria, and, as with lawns, if you have lots of earthworms and earthworm castings in your soils, then you probably have bacterially dominated soils with plenty of nitrates, which are what most vegetables and annuals prefer (remember Rule #2). Set up the Berlese funnel and see what kinds of microarthropods are roaming the soils. You want to see lots of bacteria-eating mites and good diversity of animals. Measure your soil's pH in the rhizosphere. If it is decidedly alkaline, you most probably have bacterial dominance. Similarly, an acidic reading means you have fungi and probably fungal dominance. Finally, get your soil tested for its microbiology; this is the best way to know what is missing, if anything. Sure, an NPK test won't hurt, but it is really the biology you need to know about.

No more rototilling

If you are an organic gardener you probably already employ one or two of the soil food web tools. But there is one traditional organic practice we must ask you to drop. With one exception we recommend the no-dig principle: never rototill again. This is a real shocker to those who regularly rototill or otherwise turn their soils. Soil turning is so ingrained in the psyche of the home gardener that Rule #18 is a special rule against it: rototilling and excessive soil disturbance destroy or severely damage the soil food web. They are outmoded practices and should be abandoned in established garden beds. This is heresy in most gardening circles. Many organic gardeners advocate rototilling and double digging as ways to mix organics back into the soil; indeed, rototiller manufacturers are major advertisers in magazines that promote organic gardening.

The age-old agricultural practice of plowing the earth really picked up steam, so to speak, when lawyer Jethro Tull (1674–1741) inherited a farm in southern England and invented a seed drill that mechanically placed seed at a set depth in a premade hole, replacing hand-broadcasting. Tull also actively encouraged farmers to loosen soil before planting crops; he had noticed that vegetables did better in loosened soil and from this concluded that plant roots possessed little mouths and ate soil particles (how else could a plant ingest nutrients?). Believing that loose soil consisted of smaller particles that would more easily fit into root mouths, he developed a horse-drawn hoe to put his theory into practice. His writings later caught the attention of gentlemen farmers like George Washington and Thomas Jefferson, who encouraged their fellow Americans to break up soils. The end result is that most home gardeners still break up and turn over their soil at least annually, even though we know plant roots don't eat soil.

For reasons unknown to Tull and his contemporaries, vegetables did grow better in soil that had first been loosened and to which manures were added. This had nothing to do with tiny particles of soil; it was because breaking up the soil supports Rule #2. Breaking up forest soil in order to plant a garden actually does more than make a treeless field; it reverses the results of years and years of succession, destroying the network of fungi in the soil. With fewer fungi, soils become bacterially dominant, a boon to nitrate-loving vegetables and row crops. The addition of manures by these early American farmers also greatly increased bacterial populations, as these are great bacterial food.

So, in the short term, breaking up America's virgin forest soils and mixing in manure made soils suitable for agriculture; however, rototilling or otherwise turning soil also destroys soil structure and displaces soil biota, disrupting the soil food web. It completely chops up the miles of fungal hyphae that exist even in bacterially dominated soils. Worm tunnels and the pores between soil particles are all blown apart. Sure, the soil is fluffy after rototilling, but that's a dog's name, not a soil description. The first time water hits disturbed soil, it begins to compact, a spiraling, downward course that continues every time it rains or the bed is watered.

Even bacterially dominated soils need to contain some fungi to maintain soil structure and microbial diversity. Soil food web gardening practice requires that the soil be disturbed as little as possible when it comes to annual and vegetable gardens, unless you are trying to establish a vegetable or annual garden in fungally dominated soils. Use a trowel, dowel, or dibble to make discrete holes for plants or seed. You can also lightly pull a hoe or the corner of a 2-by-4 board along a row and plant in the limited disturbed wake, backfilling with good bacterially dominated compost. You will get fewer weeds using this method because you are not opening up the soils and exposing weed seeds to the light that is required for germination.

Soil food web workers are great farmers

How do you encourage the bacterial domination needed for your annuals, vegetables, and row crops if you cannot rototill? Like everything in the soil food web, if you feed them, they shall come. Green mulches promote bacteria. In this case, not only does green mulch provide nutrients for the proper and necessary soil food web organisms, it also prevents weeds from germinating and holds moisture in, preventing it from evaporating. Too, bacteria like the easy-to-digest stuff, so the finer the green mulch, the higher the bacterial growth. Since soil bacteria also favor dampness, wetter mulches—to a point—will also promote bacteria. There is a fine line between damp, aerobic mulch and wet mulch that fosters anaerobic conditions, however, so be careful. Use your nose as the tester. If there is a bad smell, you put in too much water and need to aerate the mulch and back off a bit on the water.

In addition to bacteria-supporting mulch, your soils should have plenty of good organics to feed the microbes that are feeding your plants. Use any organic microbe food—with all three of the NPK numbers below 10—to ensure that fragile fungi are not killed. You can put these in the root zone when you plant, or side-dress before you mulch and then add as needed. Apply bacterially dominated teas as both a soil drench and foliar spray to prevent or control diseases and to keep microbial populations in the soil at high numbers.

Lawn clippings are a terrific green mulch to use around your annual flowers and your vegetables during the growing season. Even though they lose their color and turn "brown," they are still considered "green" mulch because when they were cut, they contained sugars that remain even after the chlorophyll has faded. The same is true of straw. And it is still a good idea to add organics to garden soils in autumn so they have a chance to start to break down before spring planting. Try alfalfa meal, straw, or grass clippings—all good bacteria food. The bacteria get started in autumn; during this season, they can combine all the nitrogen they need with available carbon, without interfering with any plant's needs. Nitrogen tie-up at the soil-mulch interface, if it occurs at all, will be over by spring.

When it comes to growing plants that require nitrates, good populations of protozoa and nematodes are part of the equation as well, as they are the cycling mechanism. Apply protozoa soup as a soil drench to help increase nutrient recycling in your vegetable and flower gardens. It may take a week or so for the protozoa to find the bacteria in the rhizosphere, so wash down any application of bacterial food with an immediate dose of protozoa soup. Commercial nematode products have hit the home horticultural market, but these are usually specific for garden pests such as slugs. Your best bet for increasing the populations of nutrient-cycling nematodes—and by far the most economical—is still good compost and compost tea.

Vegetable garden with straw mulch on the beds. Courtesy National Garden Bureau.

And, of course, you will have the benefit of mycorrhizal fungi working in your gardens if you follow soil food web practices. Mycorrhizae even help plants grown in containers. The longer the season, the bigger their role. This is because it takes time for these fungi to establish and grow. Rule #19 requires that soil food web gardeners always mix endomycorrhizal fungi with the seeds of annuals and vegetables at planting time or apply them to roots at transplanting time.

Of the plants that do not form mycorrhizae, many are vegetables. In particular, the families Brassicaceae (which includes cabbages, mustards, and

The potted marigolds on the right show the benefit of endomycorrhizal fungi. Courtesy Mycorrhizal Applications, www.mycorrhizae.com.

Root balls are considerably larger when corn plants (family Gramineae) are treated with endomycorrhizal fungi, as was the one shown here on the right. Courtesy Mycorrhizal Applications, www.mycorrhizae.com.

broccoli) and Chenopodiaceae (spinach, beets, lamb's-quarters) do not form mycorrhizal associations; using mycorrhizal products on these particular plants is a waste of time and money.

Once you stop applying chemicals, you will eventually find earthworms in your vegetable and flower gardens. An application of a few inches of bacterially dominated compost in early fall will help attract and support worms, as will a bacterially dominated compost tea as a soil drench. If you fail to attract earthworms, it is a sign that you need to increase bacteria and protozoa populations. Do so, and then add some worms to your annual and vegetable gardens if you want to speed things up. You can soil drench your plants once a week to once a month, depending on their performance.

Weeds

All too often the reaction of the gardener to a weed in a flower or vegetable garden is to douse it with whatever herbicide is suggested and often a bit more than the directions call for, for good measure. For obvious reasons this is not sound soil food web practice. Applications of powerful nonselective herbicides harm the soil food community in much the way chemical fertilizers do, killing micro- and macroarthropods, as well as microbes. Instead, carefully hoe weeds up or use vinegar, heat, boiling water, corn gluten, and other weed-controlling methods that have fewer and more temporary consequences to the microbiology in the soil. Should you ever need to resort to an herbicide (and we sincerely hope you won't), you must take remedial action as soon as practicable (Rule #15 again). Let the poison take its toll and then take steps, using all three soil food web tools, to get the biology back where it belongs.

When it comes to preventing weeds in the first instance, nothing beats mulches. The nitrogen, phosphate, and sulfur weeds need to germinate and grow are tied up by the biology at the interface of the mulch and the soil. This makes it doubly hard for weeds to do well, as in addition to facing no light and a physical barrier to their growth, they are given a poor supply of nutrients. Really, when you think about it, why fuss around with the other tools, compost and compost teas? Put down 2 to 3 inches (5 to 7.5 centimeters) of a bacteria-supporting mulch before weeds appear, taking care to leave a bit of "bare" soil around the stems of your plants.

Other than the work it takes to apply mulch, soil food web gardeners need never worry about weeds again. Indeed, our experience has convinced us that returning the appropriate microbiology to your soils may be the only step you'll need to control many of your annual weeds, those that thrive on the high concentrations of nitrates found in chemical fertilizers. Many of the plant pests we had in our gardens disappeared once we started working with the soil food web. Chickweed, our nemesis, completely vanished, as plants no longer got their fix of high nitrates and had trouble germinating in the first instance, their seeds buried under mulch and not exposed to light because we don't rototill.

High-nitrogen fertilizers encourage opportunistic annual weeds. Given an ample supply of nitrates, an unwanted plant suddenly has the food power to really take over. Adding to the injury, the mycorrhizal fungi your veggies and annuals use to help obtain water and nutrients, particularly phosphate, are killed. The host plant doesn't do as well; the surface-feeding, nitrate-loving weeds grow faster and overrun the garden, outcompeting the main crops for light.

Once you get the soil food web humming, any nitrates needed by plants will come from the natural course of cycling. Instead of being poured on in a concentrated, chemical form and killing off the soil food web, the only nitrates being used will be those produced by the soil food web itself. And—without chemicals and with a bit of inoculation—mycorrhizal fungi will return.

"Pests"

It is never an ideal world, unfortunately, but most insects (we use the term loosely to include spiders and others that are not truly insects) we encounter in our flower and vegetable gardens are helpful in lots of ways. Who needs to be reminded that insects pollinate flowers? Their larvae tunnel through soil and aerate it, and insects eat each other and participate in the recycling of plant nutrients. In most instances, insects get out of hand in your gardens because something is wrong with the soil food web, which normally maintains a bal-

A spined soldier bug makes a meal of the Mexican bean beetle larvae on this snap bean. Courtesy USDA-ARS.

ance between pests and predators. But you are not going to have a totally pest-free garden even with the soil food web in place. Accept it as part of the science. If your soil food web is healthy, this community will help plants overcome any insect pest. If there are a few bad guys, you need to realize that these help maintain the good-guy populations.

Every gardener has access to local agencies that will provide assistance in distinguishing beneficials from pests: learning about the beneficials in your area is part of learning to garden with the soil food web. Ladybird beetles and

A ladybird beetle larva devours aphids.
Courtesy Clemson University, USDA Cooperative Extension Slide Series, www.forestryimages.org.

A stink bug does in an eastern tent caterpillar.
Photograph by Robert L. Anderson, USDA Forest Service, www.forestryimages.org.

Braconid wasp larvae parasitize a hornworm. Courtesy R. J. Reynolds Tobacco Company, R. J. Reynolds Tobacco Company Slide Set, www.forestryimages.org.

their larvae feed on aphids, scale, and spider mites. Ground beetles eat cutworms, root maggots, slugs, and snails. Rove beetles eat fly maggots and eggs, aphids, mites, slugs, snails, and nematodes. Assassin bugs are adept at getting flies, mosquitoes, and caterpillars. Green lacewings and larvae gobble up aphids, spider mites, whiteflies, and caterpillars. Hornets take out flies. The soil food web gardener observes and learns what relationships exists—and fosters the good ones.

We don't like the use of pesticides in flower and vegetable gardens any more than we like the use of herbicides. These very nonselective substances have a flagrantly negative impact on soil food webs (again, Rule #15 will see that you rejuvenate the microbial universe in the soil and break down the residues of your action, if you have to use a pesticide). However, don't forget the lesser evils—insecticidal soaps, botanical insecticides, *Bacillus thuringiensis* (Bt)—all of which have varying impacts on the soil food web, but usually none as damaging as chemical insecticides.

Schedule for restoration and maintenance

If you habitually used chemical fertilizers in your vegetable and flower gardens, you will need all three soil food web tools. Apply 1 to 2 inches (2.5 to 5 centimeters) of bacterially dominated compost before you plant annuals and veg-

Applying compost to flower and vegetable beds. Photograph by Judith Hoersting.

etables. Spray seeds with bacterially dominated compost tea and treat them and any seedlings with mycorrhizae right before you plant them. After you plant, lay down green mulch. Start weekly applications of bacterially dominated compost tea. These measures will restore or maintain the soil food web organisms in your vegetable beds.

Spray your vegetables with a bacterially dominated compost tea as soon as the first leaves appear and at least one more time a few weeks before harvest. Spray a third application on the debris left over from the growing season.

Avoid compaction; try to stay out of the garden beds, and limit and direct pathways through them. Side-dress and top-dress plants with compost whenever possible, and put compost on garden beds before the winter. As long as it is bacterially dominated, you cannot apply too much.

Finally, it is important to mulch garden beds in autumn so that the bacteria, fungi, protozoa, and nematodes can work during the winter to cycle nutrients.

Restore and maintain the soil food webs in your flower and vegetable gardens. If we are not mistaken, the great size and taste of organically grown produce will only match the particularly lovely glow of annuals raised using the soil food web.

Chapter 21

A Simple Soil Food Web
Garden Calendar

THERE IS NO ONE WAY to garden with the soil food web. Each garden is different, and so are the various soil food webs in them. Climate, too, plays a big part in when and even how you apply soil food web science. When it is very cold, compost teas are definitely not going to work, and colder temperatures freeze up compost and mulch. Times of drought might not be the best time to apply compost tea, and putting down mulch at the wrong time in a drought situation could prevent the soil underneath from absorbing water.

Still, no matter where you garden, you should at least consider the microbes and other animals in your soil food webs as each season rolls by. Yard and garden care is no longer just about plants. You have to pay attention to the microbes if you are going to team with them.

Spring

Spring is when you first check things out and give your soils a microbiological boost. The compost pile should be cranked up so you'll have ample supply of compost throughout the growing season. Turn last fall's pile, and if you have room, start a new pile designed to be fungally dominated. Use the organic debris that accumulated during the winter and some of last fall's leaves. Use the first grass clippings to get good bacterial compost going as well.

Mulches should be pulled back to let soil warm up if necessary and then put back and supplemented. Use compost teas on seedlings both as a soil drench and a foliar spray. Inoculate all seeds and transplants with the appropriate kind of mycorrhizal fungi.

Three weeks before leaves appear, have your soils and tea tested for their microbiology. You don't have to do this every year, but you surely should the first year or two of gardening with the soil food web. Thereafter, your plants will let you know how you are doing. You might want to have your compost piles tested as well. This is also the time to test things yourself, using Berlese funnel soil traps and your own eyes. You want to be able to correct any gaps in your soil food webs before you plant.

Two weeks before leaf-out, aerate your lawns. Again this doesn't have to be done every year, but it is definitely a consideration the first year after you stop using chemical fertilizers. Thereafter, you only need to aerate in the early spring every three or four years, depending on the amount of traffic your yard receives; the amount of ice that accumulates each winter, if any; and the state of the soil food web as evidenced by worm, mite, and mushroom activity.

After aeration (or two weeks before leaf-out of trees and shrubs, if you didn't aerate), apply an appropriate organic microbe food, such as soybean meal, to lawns. If you experienced too many mushrooms (or mushrooms of only one species) the previous year, apply some alfalfa meal instead, as it will feed more bacteria than fungi.

This is also the right time to spray lawns with a slightly bacterial or balanced compost tea, at the rate of at least five gallons of tea per acre. Paths in the lawn created by winter traffic should be cordoned off and sprayed with a fungally dominated compost tea to restore structure. When you finish making teas, throw the leftover compost and any excess on these paths. After a few applications, things will be downright spongy. Even without tea, make sure the organic microbe food in these areas is sufficient to support existing microbial populations. You can't burn the lawn applying these organics, so don't worry.

Tidy up the brown mulch layer under trees and shrubs and around perennials and refresh it if you need to. This is why you should save leaves in the autumn when they drop: they can be hard to come by in the spring. If you don't have leaves, bark chips will do. You can spread compost at this time and cover it with mulch to control weeds. Apply a fungal food (humic and fulvic acids, cold-water kelps, phosphate rock dusts) to your plants, and then give each tree, perennial, and shrub a soil drench of your most fungal compost tea. Spray a fungal tea on your perennials at least once after their leaves appear.

Treat any seeds or transplants with the appropriate mycorrhizal fungi first. If possible soak transplants in aerated compost tea before planting. Spray compost tea on seeds before planting, and apply a soil drench after germination.

Neither till the vegetable garden nor turn over the soil in the annual beds. Apply 4 lbs per 100 square feet soybean meal as soon as you can after the soils thaw, and spray with a bacterially dominated tea. When planting, drill holes for seeds or disturb just the row where they will be planted. Use lots of green mulch after the soil warms up.

Summer

During the summer months you need to continue with the spray and drench program started in the spring, especially the first year after you stop using chemicals.

Microbial activity should be taking care of lawn clippings. If they are accumulating at a noticeable rate, or the lawn is not greening up enough and lack of water is not the cause, spray or sprinkle on a protozoa soup. A second application of soybean meal or other microbe food is in order. It is useful to do more Berlese funnel tests to see what is going on. Keep records for later comparisons.

Liberal applications of bacterially dominated compost and frequent replenishment of green mulches will keep weeds down in vegetable and annual gardens. Apply microbe food once every two weeks if needed.

Fungal compost and mulch should be applied liberally around trees, shrubs, and perennials. Mix in any twigs or sticks these plants drop. You might run these over with a lawn mower, in place, just to open them up a bit and make them look neater.

Any plants showing signs of disease or stress should be immediately sprayed with compost tea followed by a soil drench of tea.

Autumn

Just before the tree leaves start to drop, gather up a load of grass clippings for fall composting, which should begin while the grass is still fresh and green. You can also put some of this green mulch on annual and vegetable beds, even if the season is coming to an end. Use mycorrhizal fungi on the roots of any autumn transplants.

Turn leaves that fall on lawns into a fine mulch with your mower (you may have to run over them more than once). Leave them in place. This will provide some fungal balance to the bacterially dominated teas you have been applying to the lawn. Gather the rest of the leaves, every single one you can. Brown leaves are always in short supply when it comes to spring and summer composting. Build your compost pile and store the rest.

Mulch vegetable and flower garden beds. After leaf drop, make sure all your shrubs, trees, and perennial plants are properly mulched, too, and if possible, use fungally dominated compost first.

In the first year of using the soil food web, spray 20 gallons of tea per acre, making sure to inoculate mulches and leaves. Microbial action should decay about half the leaf mass within a month or so if it is warm (and by the end of spring, even if it is cool).

Apply a good organic microbial food of the appropriate type. Let the microbes go to sleep with full stomachs, wake up early, and start cycling nutrients.

After harvests, have your soils tested again and make some Berlese funnel runs, if it is not too cold; compare these tests to those you took in the spring and summer. This will allow you to manage your soils during the winter months so they are ready come next spring.

Winter

Spend winter reading up on the soil food web, surfing the Internet and browsing libraries with that subject in mind. This is a new science, and its applications to the home gardener are ever expanding. New products, such as specialized predatory bacteria and nematodes that take out pests and pathogens, are being introduced all the time. All sorts of new compost tea makers, sprayers, and nutrient ingredients are hitting the market. There is a lot out there to help you team with microbes, and you need to keep abreast of the latest developments.

Of course, just because it's winter doesn't mean you should stop using compost teas. You can have an abbreviated soil food web system working for your indoor plants; make sure the potting soils contain ample organic foods to support the microbial life you are adding.

Finally, depending on where you live, your compost pile may still be workable in the winter. Give it a few turns. You know the saying: a few good turns will make you a better gardener.

Chapter 22

No One Ever Fertilized an Old Growth Forest

DOES THE SOIL FOOD WEB really support plants? Will it work in your yard and gardens? Just to give you confidence and to encourage you to use what you have learned, we point you in the direction of the nearest forest. Or simply close your eyes and visualize any wooded area you remember visiting. You can almost hear a stream nearby, the wind running through the leaves. It is beautiful, majestic—and no one ever fertilized any of the plants there. Not one single time. How can this be? You know the answer. The beautiful plants in these beautiful areas are completely controlled by the soil food webs in which they live.

It often comes as a surprise when gardeners so reflect. Only then does the full force of the realization hit: every single plant you are seeing produces exudates and attracts microbiology to its rhizosphere. This community in turn attracts micro- and macroarthropods, worms, mollusks, and the rest of a complete soil food web. It is a natural system, and it operates just fine without interference from man-made fertilizers, herbicides, and pesticides. Tall oaks grow from small acorns with no blue powders to feed them or nasty smelling sprays to protect them. Plants flourish nonetheless, thanks to bacteria, fungi, protozoa, nematodes, and the rest of the soil food web gang.

We know it is possible to let the very same kind of soil food webs take over in your yard. Long before construction, traffic, rototilling, the application of fertilizers and other chemicals, a healthy soil food web existed there. You can return it. You can even improve it. Once you work with the microbes at the base of the soil food web, you will reestablish that soil food web. We know. We and thousands of our neighbors and friends have done it.

You have been introduced to the basic science of soil food webs. You know how the system works, and you have been exposed to its benefits. With the microbiology returned to your yard, soil structure improves. Mycorrhizal fungi will help your lawn, trees, shrubs, perennials, annuals, and veggies get the nutrients they need. Pathogens face fierce competition. Plants get more of the kind of nitrogen they prefer. Water drainage and retention are improved. Pollutants are decayed. Food tastes better. Flowers look better. Trees are less stressed. And you don't have to work so hard; you will have lots of helpers. Best of all, you won't have to worry about the affects of chemicals on you or your family, pets, or friends.

Remember: no one ever fertilized an old growth forest. They didn't have to. You have been given the rules to garden using the soil food web. There are not many of them. What are you waiting for? Start teaming with microbes and get that biology into your soils and working for you. Gardening with the soil food web is the natural way to grow.

No one fertilized this forest. Photograph by Judith Hoersting.

Appendix

The Soil Food Web
Gardening Rules

1. Some plants prefer soils dominated by fungi; others prefer soils dominated by bacteria.

2. Most vegetables, annuals, and grasses prefer their nitrogen in nitrate form and do best in bacterially dominated soils.

3. Most trees, shrubs, and perennials prefer their nitrogen in ammonium form and do best in fungally dominated soils.

4. Compost can be used to inoculate beneficial microbes and life into soils around your yard and introduce, maintain, or alter the soil food web in a particular area.

5. Adding compost and its soil food web to the surface of the soil will inoculate the soil with the same soil food web.

6. Aged, brown organic materials support fungi; fresh, green organic materials support bacteria.

7. Mulch laid on the surface tends to support fungi; mulch worked into the soil tends to support bacteria.

8. If you wet and grind mulch thoroughly, it speeds up bacterial colonization.

9. Coarse, dryer mulches support fungal activity.

10. Sugars help bacteria multiply and grow; kelp, humic and fulvic acids, and phosphate rock dusts help fungi grow.

11. By choosing the compost you begin with and what nutrients you add to it, you can make teas that are heavily fungal, bacterially dominated, or balanced.

12. Compost teas are very sensitive to chlorine and preservatives in the brewing water and ingredients.

13. Applications of synthetic fertilizers kill off most or all of the soil food web microbes.

14. Stay away from additives that have high NPK numbers.

15. Follow any chemical spraying or soil drenching with an application of compost tea.

16. Most conifers and hardwood trees (birch, oak, beech, hickory) form mycorrhizae with ectomycorrhizal fungi.

17. Most vegetables, annuals, grasses, shrubs, softwood trees, and perennials form mycorrhizae with endomycorrhizal fungi.

18. Rototilling and excessive soil disturbance destroy or severely damage the soil food web.

19. Always mix endomycorrhizal fungi with the seeds of annuals and vegetables at planting time or apply them to roots at transplanting time.

Resources

American Phytopathological Society. "Plant Pathology on Line." http://www.apsnet.org/education/K-12PlantPathways/Top.html.

————. "Illustrated Glossary of Plant Pathology." http://www.apsnet.org/education/IllustratedGlossary/default.htm.

BioCycle. The JG Press, Inc., 419 State Ave., Emmaus, PA 18049, 610.967.4135, biocycle@jgpress.com, http://www.jgpress.com/biocycle.htm.

Bugwood Network, USDA Forest Service / University of Georgia, Warnell School of Forest Resources and College of Agricultural and Environmental Sciences, Dept. of Entomology. "Forestry Images." www.forestryimages.org.

Carroll, S. B., and S. D. Salt. 2004. *Ecology for Gardeners*. Timber Press: Portland, Ore.

Cloyd, R. A., et al. 2004. *IPM for Gardeners*. Timber Press: Portland, Ore.

Dennis Kunkel Microscopy, Inc. "Science Stock Photography." http://denniskunkel.com/.

Grissell, E. 2001. *Insects and Gardens.*Timber Press: Portland, Ore.

Hall, I., et al. 2003. *Edible and Poisonous Mushrooms of the World*. Timber Press: Portland, Ore.

Helyer, N., et al. 2003. *A Color Handbook of Biological Control in Plant Protection*. Timber Press: Portland, Ore.

Ingham, E., et al. 2000. *Soil Biology Primer*. Soil & Water Conservation Society and USDA Natural Resources Conservation Service, 7515 NE Ankeny Rd., Ankey, IA 50021-9764, http://www.swcs.org.

Kilham, K. 1994. *Soil Ecology*. Cambridge University Press: London.

McBride, M. B. 1994. *Environmental Chemistry of Soils.* Oxford University Press: New York.

Paul, E. A., and F. E. Clark. 1989. *Soil Microbiology and Biochemistry.* Academic Press: San Diego.

Stephenson, S. L., and H. Stempen. 1994. *Myxomycetes: A Handbook of Slime Molds.* Timber Press: Portland, Ore.

Sylvia, D. M., et al. 1998. *Principles and Applications of Soil Microbiology.* Prentice Hall: Upper Saddle River, N.J.

United States Department of Agriculture, National Resources Conservation Services. "Soil Quality." www.forestryimages.org.

————, Agricultural Research Service. "Online Photo Gallery and Photo Library Archives." Conservation Communications Staff, Box 2890, Washington, DC 20013, http://www.ars.usda.gov/is/graphics/photos/search.htm.

United States Department of Interior, Bureau of Land Management. "Soil Biological Communities." National Science and Technology Center, Box 25047, Bldg. 50, Denver Federal Center, Denver, CO 80225-0047, 303.236.2772, http://www.blm.gov/nstc/soil/.

Weeden, C. R., et al., eds. "Biological Control: A Guide to Natural Enemies in North America." Cornell University. http://www.nysaes.cornell.edu/ent/biocontrol/.

White, D. 1995. *The Physiology and Biochemistry of Prokaryotes.* Oxford University Press: New York.

Worm Digest. Worm Forum, Box 544, Eugene, OR 97440-0544, mail@wormdigest.org, http://www.wormdigest.org/forum/index.cgi.

Composting and compost tea

California Integrated Waste Management Board. "Compost Microbiology and the Soil Food Web." http://www.ciwmb.ca.gov/publications/Organics/44200013.doc.

Composting News. McEntee Media Corp, 13727 Holland Rd., Cleveland, OH 44142, 216.362.7979, mcenteemedia@compuserve.com, http://www.recycle.cc/cnpage.htm.

Compost Science and Utilization. The JG Press, Inc., 19 State Ave., Emmaus, PA 18049, 610.967.4135, biocycle@jgpress.com, http://www.jgpress.com/compost.htm.

Compost Tea Forum. http://groups.yahoo.com/group/compost_tea/.

Cornell University. "Cornell Composting." http://compost.css.cornell.edu/Composting_homepage.html.

Diver, S. 2002. "Notes on Compost Teas." Appropriate Technology Transfer for Rural Areas (ATTRA). http://attra.ncat.org/attra-pub/compost-tea-notes.html.

Granatstein, D. 1997. "Suppressing Plant Diseases with Compost." *The Compost Connection for Washington Agriculture* 5 (October). http://csanr.wsu.edu/programs/compost/Cc5.pdf.

Ingham, E. 2000. *The Compost Tea Brewing Manual.* Soil Food Web, Inc. Corvallis, Ore. http://www.soilfoodweb.com.

————. 2000. "Brewering Compost Tea." *Kitchen Gardener* 29 (October). http://www.taunton.com/finegardening/pages/g00030.asp.

————. 2004. *Compost Tea Quality: Light Microscope Methods.* Soil Food Web, Inc. Corvallis, Ore. http://www.soilfoodweb.com.

————. 2004. *The Field Guide to Actively Aerated Compost Tea.* Soil Food Web, Inc. Corvallis, Ore. http://www.soilfoodweb.com.

International Compost Tea Council. http://www.intlctc.org/default.asp.

Large-Scale Composting Forum. http://www.oldgrowth.org/compost/forum_large/index.html.

Ringer, C. "Bibliography on Compost for Disease Suppression." USDA Soil Microbial Lab. http://ncatark.uark.edu/~steved/compost-disease-biblio.html/.

Tranker, A., and W. Brinton. "Compost Practices for Control of Grape Powdery Mildew (*Uncinula necator*)." A *Biodynamics Journal* reprint. http://www.woodsend.org/will2.pdf.

Vermicompost Forum. http://www.oldgrowth.org/compost/forum_vermi1/.

Compost tea brewers

Alaska Bountea / Alaska Bounty, Box 1072, Palmer, AK 99645, 907.745.8234, order@alaskagiant.com, http://www.alaskagiant.com/

Bob's Brewers, 6515 W. Marginal Way SW, Seattle, WA 98106, 206.767.7816, bob@bobsbrewers.com, http://www.bobsbrewers.com/

Keep It Simple (KIS), Inc., 2323 180th Ave. NE, Redmond, WA 98052-2212, 866.558.0990, kis@simplici-tea.com, www.simplici-tea.com, www.kisbrewer.com

Soil Soup, 305 9th Ave. N, Seattle, WA 98109, 877.711.7687, www.soilsoup.com

Labs that perform biological testing

AgriEnergy Resources, 21417 1950 E. St., Princeton, IL 61356, 818.872.1190, info@agrienergy.net, http://www.agrienergy.net/.

BBC Laboratories, Inc., 1217 N. Stadem Dr., Tempe, AZ 85281, 480.967.5931, bbclabs@aol.com, http://bbclabs.com/.

Soil Foodweb, Inc., 728 SW Wake Robin Ave., Corvallis, OR 97333, 541.752.5066, sfi@soilfoodweb.com, http://www.soilfoodweb.com.

Mycorrhizal fungi

Mycorrhizal Applications, Inc., Box 1181, Grants Pass, OR 97528, 866.476.7800, http://www.mycorrhizae.com/index.php?cid=60.

T and J Enterprises, Thomas Giannou, 2328 W. Providence Ave., Spokane, WA 99205, 509.327.7670, http://www.tandjenterprises.com/.

Index

Boldface page ranges indicate the main discussion of a topic.